T0257745

Engineering Management

Engineering Management

Edited by **Ruth Hinrichs**

LANRYE
INTERNATIONAL

New Jersey

Published by Clanrye International,
55 Van Reypen Street,
Jersey City, NJ 07306, USA
www.clanryeinternational.com

Engineering Management
Edited by Ruth Hinrichs

International Standard Book Number: 978-1-63240-210-3 (Hardback)

Printed in the United States of America.

Contents

Preface

In my initial years as a student, I used to run to the library at every possible instance to grab a book and learn something new. Books were my primary source of knowledge and I would not have come such a long way without all that I learnt from them. Thus, when I was approached to edit this book; I became understandably nostalgic. It was an absolute honor to be considered worthy of guiding the current generation as well as those to come. I put all my knowledge and hard work into making this book most beneficial for its readers.

This book combines engineering principles with business practice, i.e. it gives a consolidation of the primary fields of engineering and technology with the organizational, administrative, and planning capacities of management. It corresponds to various other fields like finance, marketing, economics, among others. The contributions in this book demonstrate the original work done in this area and give case studies which have successfully applied engineering management in real life situations. It will be beneficial for readers researching new developments in this field or for those utilizing this field as part of their work.

I wish to thank my publisher for supporting me at every step. I would also like to thank all the authors who have contributed their researches in this book. I hope this book will be a valuable contribution to the progress of the field.

Editor

Heuristic Approaches for a Dual Optimization Problem

Fausto Pedro García Márquez and
Marta Ramos Martín Nieto

Additional information is available at the end of the chapter

1. Introduction

The current crisis in the global economy and the stiff competition has led many firms to recognize the importance of managing their logistic network for organizational effectiveness, improved customer value, better utilization of resources, and increased profitability. The logistics business in Spain continues rising mainly by the new electronics market. In 2008 the turnover of logistics activities was 3.745 m€, 1.5% compared to previous year. Despite the upward trend, the strategic sector analysis done by DBK shows that the problem of declining business performance of the sector is as a result of rising fuel prices. The same study claim that the industry is in a process of concentration, with the disappearance of small operators (DBK (2009)). This requires that firms need to optimize its efficiency, e.g. recalculating the routes in order to minimize costs. To reduce the logistics costs related to transportation routes is a goal sought by all firms, where the transportation costs are easily controlled in the value chain.

There is a difference between national and international transport by road, and the distribution within the city and its close environment (widespread distribution). It has been more important in nowadays, where many firms need to do their deliveries at close proximity. However, when transportation at national and international levels is involved, more benefits can be achieved by a good planning strategy. The national and international transport by road, e.g. transport between urban centres, requires large vehicles carrying its maximum load. A good route planning can reduce the costs significantly, especially when the increasing in oil prices makes any unnecessary kilometre a profit to the company.

In Spain there are approximately 225 logistic firms, but only 4 of them have the majority of market (more than 40% of the total). In this study the biggest one, with 3577 vehicles and 411 great vehicles, has been considered. The company is focused on the distribution into cities by road. The routes are interconnected through ships, i.e. a high capacity logistic centres that are strategically located. The company has designed its domestic routes based on its own experience. This paper presents a meta-heuristic method that determines the routes that involve the major number of cities in order to increase flexibility, leading the vehicle deliver and pick up in these cities, trying to minimize the distance travelled and its costs.

The problem can be approximated to a series of problems similar to the travel salesman problems (TSP), and therefore a vehicle routing problem, with time windows (VRPTW). VRPTW tries to find a final solution including sub-path not connected and that meet the constraints of the TSP considering the time windows constrains. The restrictions of Miller et al. (1960) have been used in order to reduce the computational cost. The main purposes of the method are to provide a quick solution and flexible enough to be used in a dynamic scheduling environment, and to develop a new solution procedure that is capable of exploiting the special characteristics of the problem.

Drawing upon the state of the art presented in next section is developed a recurrent neural network approach, which involves not just unsupervised learning to train neurons, but an integrated approach where Genetic Algorithm is utilized for training neurons so as to obtain a model with the least error.

The paper is organized as follows. Section 2 elaborates on the problem faced along with the considered case study. Section 3 describes TSP modelling for the problem and a brief state of the art on VRP and applied heuristics. Section 4 provides the working of heuristics and computational experience for the recurrent neural network approach and genetic algorithm, and finally Sections 5 and 6 explain the results and conclusion respectively.

2. Case study

2.1. Background

The main problem is the profitability of routes for the logistic companies. This research paper analyses cases where a direct route between two cities minimize the distances, but it should not be considered from the cost point of view because the shipping volume is not significant enough. On the other hand, with a significant shipment, it is economically more profitable that when only the distance between two cities are considered. Therefore the transport logistics should be designed considering the service effectiveness.

The main objective is to satisfy the customers with greater effectiveness and efficiency, especially with the competence. Routes are constructed to dispatch a fleet of homogenous or heterogeneous vehicles to service a set of customers from a single distribution depot. Each vehicle has a fixed capacity and each customer has a known demand that must be fully satisfied. The objective is to provide each vehicle with a route that maximize the cities visited and the total distance travelled by the fleet (or the total travel cost incurred by the fleet), minimising the costs.

The problem is characterized as follows: From a principal depot the products must be delivered in given quantities to certain customers. A number of vehicles with different capacities are available. All the vehicles that are employed in the solution must cover a route, starting and ending at the principal depot, and the products are delivered to one or more customers in the route. The problem consists in determining the allocation of the customers among routes and the sequence in which the customers shall be visited on a route. The objective is to find a solution which minimizes the total transportation costs. Furthermore, the solution must satisfy the restrictions that every customer is visited exactly once in the capillary routes where the demanded quantities are delivered, but it is not necessary for the principal route. The transportation costs are specified, where the costs are not necessarily identical in the two directions between two cities.

In this paper the meta-heuristics method of the recurrent neural network is proposed to solve the dual problem, in order to increase the flexibility in the routes and minimizing costs. The following considerations have been considered:

- The principal route will be covered by a large-capacity truck. For practical purposes, it will be considered a big commercial vehicle.

- Capillary routes (routes between a principal city and the near small cities) will be covered by trucks of medium / small capacities, considered a light commercial vehicle.

- The First-Input-First-Output (FIFO) method is followed when multiple vehicles are present to transhipment transport.

- The fuel consumption is taken as an average value of 30 litres per 100 km for a big vehicle, and 15 litres per 100 km for a light commercial vehicle.

- The diesel price is fixed as 1 € / litre.

- The maximum speed considered are the legally permissible for a vehicle of these characteristics according to the Spanish laws.

2.2. Principal and capillary routes

A real case study has been considered, which the principal route consists in determining the route for sending a product set from Barcelona to Toledo (Spain). The route considered by the company is:

Main Route 1: Barcelona-Madrid; Main Route 2: Madrid-Toledo; Capillary route: Toledo-different towns close to Toledo.

The Madrid-Barcelona route is the same to the Barcelona-Madrid. The total distance is 1223 km, and the time estimated is 14 hours and 41 minutes, with a total cost of 366.93 €.

A first approach in this case study is to employ a route which passes through the maximum number of cities as possible minimizing costs, with the objective of maximise the flexibility. It will lead to the vehicle pick up or deliver products in those cities.

A big capacity vehicle covers this route, denoted as 'vehicle A', leaving the origin city with a certain quantity of product. If there is excess of products, they will be transported by other vehicles.

The solution proposed by the company is: Vehicles follow the route assigned to arrive in Madrid. The vehicles are unloaded and are available to be loaded again. The vehicles leave for Barcelona and the availability of products in order to fill the vehicles is not assured. The vehicle A must to wait to be fill, creating waiting time that increases the logistic costs. The vehicle A can be then loaded for shipment to Cuenca (an intermediate city), and other trucks that make the route Cuenca-Madrid-Cuenca are unloaded in Madrid.

The vehicle A will serve as a logistical support, which means that normally it will be loaded partially. It will be loaded completely in Teruel (city in the middle of the route). The products will be unloaded in Cuenca, first destination from Madrid, and then loaded with new products to be shipped in Teruel, next destination before to arrive to Barcelona, last destination. The same process followed for the city of Cuenca is applicable to the city of Teruel as it has been abovementioned.

This procedure done by the logistic company justifies the need of visiting the maximum number of logistic cities in any route. But if the vehicle visits many cities appears delay problems or the increasing of the costs.

When the vehicle arrives from Madrid to Toledo (a direct route that will not be considered in the dual problem), the products require to be served in different towns close to Toledo. It is done following capillary routes.

The case study considers a new vehicle that visits ten towns, starting and finishing in Toledo. In any town that is visited the vehicle need to deliver and to pick up products according to the orders processed in the previous day (for delivery) or in any specific day (in the case of pick up). The assigned route by the company is:

Toledo → Torrijos → Bargas → Mocejón → Añover de Tajo → Recas → Yuncos → Illescas → Esquivias → Fuensalida → Toledo.

The total distance covered is 209.9 km, and the time is 2 h 41 min.

In this paper a solution is found out for the dual problem, maximizing the logistic centres visited and minimizing the distance covered, considering the restrictions of the current time and costs given by the company.

3. Dual problem formulation

3.1. Travelling salesman problem (TSP) approach for the primary distribution

TSP consists in finding a route with the shortest distance that visit all the nodes (cities) and only once each, starting in a city and returning to the starting city (Nilsson, 1982). TSP has been very important because the algorithms developed to solve it do not guarantee to solve it with optimality within reasonable computational cost. Therefore a great number of heuristics and heuristics algorithms have been developed to solve this problem in approximately form. TSP is a NP-hard problem in combinatorial optimization that requires finding a shortest Hamiltonian tour on n given cities (Lawler et al. 1985; Gutin and Punnen 2002). Cities are represented by nodes in a graph, or by points in the Euclidean plane. The distances between n cities are stored in a distance matrix $\mathbf{D}$ with elements d_{ij}, being d_{ij} the distance between cities i and j, where the diagonal elements d_{ii} are zero, i.e. there is not distance between a city and itself. A common assumption is that the triangle inequality holds, that is $d_{ij} \leq d_{ik} + d_{kj}$, $\forall \ i,j,k = 1,\ldots,n$. Also, the symmetrical assumption, $d_{ij}=d_{ji}$, it is the same distance from i to j than from j to i. A review of previous works on TSP using different heuristics is provided in Table 1.

Methods	References
Branch-and-bound	Finke et al. (1984); Little et al. (1989); Balas and Toth (1985); Miller and Pekny (1989)
2-opt	Lin and Kernighan (1973); Bianchessi and Righini (2007); Gendreau et al. (1999); Potvin et al. (1996); Tarantilis (2005); Tarantilis and Kiranoudis (2007); Verhoeven et al. (1995)
Insertion	Breedam (2001); Chao (2002); Daniels et al. (1998); Lau et al. (2003); Nanry and Barnes (2000)
Neural network	Shirrish et al.(1993); Burke (1994)
Simulated annealing	Kirkpatrick et al.(1985); Malek et al. (1989); Osman (1993)
Tabu search	Glover (1990); Gendreau et al. (1996); Gendreau et al. (1998); Ahr and Reinelt (2006); Augerat et al. (1998); Badeau et al. (1997); Brandao and Mercer (1997); Barbarosoglu and Ozgur (1999); Garcia et al. (1994); Semet and Taillard (1993); Hertz et al. (2000); Montane and Galvao (2006); Scheuerer (2006)
Exact methods	Carpaneto and Toth (1980); Fischetti and Toth (1989); Gouveia and Pires (1999); Lysgaard (1999); Wong (1980)
Genetic Algorithm	Gen and Cheng (1997); Potvin (1996); Moon et al.(2002)

Table 1. Literature summary: different heuristic methods for solving TSP.

The heuristics algorithms developed for solving the TSP presents low computational cost and provides solutions near to the optimal. Different approaches have leaded different formulations for solving the TSP as a linear programming problem, with integer/mixed integer variables (Lawler et al., 1985, and Junger et al., 1997). Many managerial problems, like routing problems, facility location problems, scheduling problems, network design problems, can be modelled as TSP. A great number of articles have appeared with detailed literature reviews for TSP, e.g. Bellmore and Nemhauser (1968), Bodin (1975), Golden et al. (1975), Gillett and Miller (1974), and Turner et al. (1974).

The problem presented in this paper is formulated as a TSP approach for the principal distribution with the travel cycle known as a Hamiltonian cycle, i.e. the problem is defined by the graph $\mathbf{G} = (\mathbf{V}, \mathbf{E})$, where $\mathbf{V} \in \mathfrak{R}^2$ is a set of n cities, and $\mathbf{E}$ is a set of arcs connecting these cities, but in this approach the cities can be visited more than once. Under these conditions, the problem can be formulated as:

Minimize:

$$\sum_{i<k} c_{ij} x_{ij},$$

(1)

where x_{ij} is the binary decision variable that when $i<j$ has the following values:

$$x_{ij} \begin{cases} 1 & \text{if the arc joining cities i and j is used in solution} \\ 0 & \text{otherwise} \end{cases},$$

subject to the constraints:

$$\sum_{i<k} x_{ik} + \sum_{j<k} x_{kj} = 2,$$

(2)

$$k = 1, 2, \dots n,$$

(3)

$$\sum_{i,j \in S} x_{ij} \le |S| - 1,$$

(4)

$$S \subset V, \quad 3 \le |S| \le n-3, \quad x_{ij} \in \{0,1\},$$
$$i, j = 1, 2, \dots n, \quad i < j,$$

(5)

being equation 1 the objective function. C is the associated cost matrix to the matrix **E**, compounds by the elements c_{ij} that represents the "distance" (expressed in physical distance, cost, time, etc.) between the cities i and j, where $c_{ij} \le c_{ik} + c_{kj}$ for all $i,j \in V$, to be Euclidean. The constraints ensure that:

i. All cities are connected to each other

ii. Elimination of sub-path S since the sub-path should not be defined for $|S|=2$ and $|n-2|$ because restrictions (iii) and (iv) ensure that between two cities no sub-path is generated.

The model (1) contains n $(n-1)$ binary variables, with $2n$ constraints and $2^n - 2(n-1)$ sub-path constraints that need to be removed, making it very complex and computational costly. The restrictions proposed by Miller et al. (1960) have been considered which can reduce the number of sub-path, also referred to as disposal restrictions. In these new restrictions is necessary to consider the new variables u_i $(i = 2,\dots, n)$ given by:

$$u_i - u_j + (n-1)x_{ij} \leq n-2, \quad i,j = 2,...n, \quad i \neq j, \tag{6}$$

$$1 \leq u_i \leq n-1, \quad i = 2,...n. \tag{7}$$

The restriction (1.v) indicates that the solution does not contain a sub-path in all cities $S \subseteq V$ and all sub-path contains more than n cities. The restriction (1.vi) ensures that the u_i variables are defined only for each sub-path. This formulation has been employed for solving the principal distribution, e.g. the transport between the cities of Barcelona and Madrid, considering the main cities between them, where it is possible to visit a city more than once.

TSPs can also be represented as integer and linear programming problems. In this paper it will employed for the capillary formulation problem. The integer programming (IP) formulation is based on the assignment problem with additional constraint of no sub-tours:

$$\min \sum_{i=1}^{n} \sum_{j=1}^{n} c_{ij} x_{ij} \tag{8}$$

$$\text{s.t. } \sum_{i=1}^{n} x_{ij} = 1 \text{ for all } j \tag{9}$$

$$\sum_{j=1}^{n} x_{ij} = 1 \text{ for all } i \tag{10}$$
$$x_{ij} \in \{0,1\} \text{ for all } i,j$$

where (2) is the objective function and the constraints (3) and (4) ensure that each city is visited exactly once. TSP can be also expressed as a linear programming (LP) formulation by the equation (5).

$$\text{Minimize } \sum_{i=1}^{m} w_i x_i = w^T x \tag{11}$$
$$\text{Subject to } \quad x \in S$$

where m is the number of edges in G, w_i is the weight of edge and x is the incidence vector that indicates the presence or absence of each edge in the tour. There are a number of algorithms used to find optimal tours, but none are feasible for large instances since they all grow exponentially. This formation has been employed for solving the capillary route problem.

3.2. Vehicle route problem

The transport problem is formulated in this paper as the travel salesman problem (TSP) adapted to the real case study, adding different routes to the final route. Therefore it will be considered as a VRP problem.

The VRP has been considered in many research works in the last few years. Evans and Norback (1985) designed an heuristic-based decision-support system, which utilizes computer-graphic pictures of routes in a large service distribution. The system provides scheduler of routes with a tool to enable the rapid evaluation of computer-proposed solutions and to easily modify them.

Faulin (2003) employed the MIXALG method, combining heuristic algorithms and linear programming routine, as a way of solving routing problems with moderated size. This method is efficient because does not consider some burdensome procedures in unnecessary situations. The initial solutions for linear programming have been found by a Clarke–Wright variant method that considers the logistic cost reduction as one of the main conclusions.

Hsu and Feng (2003) studied the distribution using a VRP with time windows (VRPTW), and they solved the problem by the Time-Oriented Nearest-Neighbor Heuristic method. Huey-Kuo et al (2009) employed a nonlinear mathematical model, based on the constrained Nelder–Mead method and a heuristic algorithm, for a VRPTW, with the objective of maximising the expected total profit of the supplier setting the optimal production quantities, the time to start producing and the vehicle routes. Loannou et al. (2001) solved the VRPTW using a heuristic method based upon Atkinson's greedy look-ahead heuristic.

Ma et al. (2012) solved a vehicle routing problem with time windows and link capacity constraints (VRPTWLC). They employed a tabu search heuristic with an adaptive penalty mechanism (TSAP).

Prins (2004), contrary to the VRPTW, concludes that no genetic algorithm can compete with the tabu search (TS) methods designed for the VRP. Prindezis et al. (2003) developed an Application Service Provider to coordinate and disseminate tasks and related spatial and non-spatial information for solving the VRP. A similar case was solved by Gendrau et al. (2006) employing TS. In 2004, Tarantilis et al. employing a metaheuristic algorithm called BoneRoute, for solving the open vehicle routing problem (OVRP). The OVRP deals with the VRP problem without returning to the distribution centre. The VRP with backhauls (VRPB), where deliveries after pickups are not allowed is solved by Tütüncü et al. (2009). The authors extended the formulation to a mixed VRPB where deliveries after pickups are allowed. A new criterion, which considers the remaining capacity of the vehicles, is proposed to find solutions for mixed and restricted VRPB. They solved the problem a greedy randomised adaptive memory programming search (GRAMPS) algorithm.

The problem was formulated by Tarantilis and Kiranoudis (2001) as an open multi-depot vehicle routing problem (OMDVRP). It was solved by a stochastic search meta-heuristic algorithm termed as the list-based threshold accepting (LBTA) algorithm. The proposed routing plan gives answers to a number of operational decision problems and provides significant economic benefits for the company. An extension of this study was presented in Tarantilis and Kiranoudis (2002).

4. Recurrent neural network and genetic algorithm approaches (RNNGA)

Neural networks (NN) represent the operating mechanism of the human brain, based on a fair degree of some simple computational nodes called neurons. The knowledge is acquired through a learning process, and the connection interneuron (synaptic weights) would be used for the storage of knowledge. Artificial NN are networks comprising of large quantities of highly interconnected simple computational elements. They use data from previous steps incorporating information from multiple indicators, being a non-parametric model (Alekxander and Morton, 1990). Time and data are required for learning and training the network. Once the network is trained and completed, it can determine feasible solutions to similar problems. Figure 1 shows the structure of a NN where each neuron receives information from neurons that are found in a layer closer to the input layer, and sends the output to a layer that is closer to the output layer. The types of links in the NN consist of synaptic and activation links, and the way in which neurons in the network structure are assigned determines its architecture. NN are non-linear statistical data modelling tools used to model complex relationships between inputs and outputs or to find patterns in data. Recurrent Neural Network (RNN) refers to a special type of neural network where the output of previous iteration is used as an input for the next iteration. There are many systems in the real world whose behaviour depends on their current state, such systems can be modelled by RNN. When the NN is applied to problems involving nonlinear dynamical or state dependent systems, NN with feedbacks can in some cases provide significant advantages over purely feed forward neural network (FNN).

There are some input neurons and one feedback neuron. The feedback neuron takes previous iteration's output as input while the other neurons take a fixed input. The output of the input layer is passed to hidden layer; output of interaction of hidden layer neurons is passed to the output, therefore it gets an output. The associated weights are calculated by applying some algorithm, e.g. back propagation using gradients. In this research work a genetic algorithm (GA) is used to determine the weights.

Back propagation method using gradients for training has been successfully applied to FNN (Bourlard and Wellekens 1989, Le Cun et al. 1989, Sejnowski and Rosenberg 1987, Waibel et al. 1989). However this training algorithm has not been successful for recurrent NN due to complexities (Blanco et al. 1990). Training algorithms for RNN, based on the error gradient, are very unstable in their search for a minimum and require much computational time when the number of neurons is high (Blanco et al.2000). This is the main reason where it is proposed a GA to evaluate weights.

The fitness function error is calculated as follow: Firstly, the weights in the network are set according to the weight vector; then the network is evaluated against the training sequence. It will lead to determinate the sum-squared-difference between training sequence and the known target values employed in the training sequence in each vector. The GA is adjusted to the weights, being the network represented by a chromosome and the weight link in summarised in one gene. There are many chromosomes that make up the population, therefore, many different neural networks are evolved until the minimum value of the mean-squared-error is satisfied. The fitness function evaluates the mean squared error in the training process for each NN, being the main objective to minimise the function.

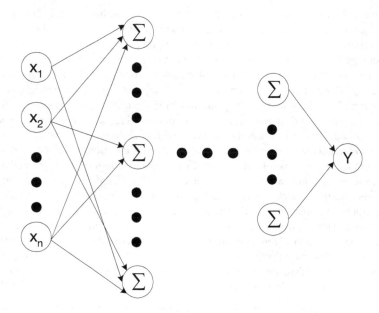

Figure 1. The system structure of a recurrent neural network

The output of the network can now be represented as:

$$\mathbf{Y}(t) = f\left(\sum_{j=1}^{N} f\left(\sum_{i=1}^{10} U_{ij}\mathbf{X}_i(t)\right)W_j\right) \qquad (12)$$

$$\mathbf{X}_{10}(t) = \mathbf{Y}(t-1) \qquad (13)$$

$$\mathbf{X}_{10}(0) = 0 \qquad (14)$$

where Y(t)= Output in iteration t. $\mathbf{X}_i(t)$= Input i at iteration t. U_{ij}= Weights between input and the hidden layer. W_j= Weights between hidden layer and the output node. f = Activation function.

N is the number of neurons in the hidden layer.

The nomenclature followed is that U_{ij} connects jth node in input layer to ith node in hidden layer, similarly for W_j.

Let d be the desired output for k^{th} input, the error will be

$$Z_k(t) = |Y_k(t) - d_k| \tag{15}$$

The objective function for the GA is Z, which is the mean of square of errors for all values of inputs of X_j's. Mathematically

$$Z(t) = \sum_{k=1}^{n} Z_k(t) * Z_k(t) / n \tag{16}$$

The steps for GA employed are summarized in Figure 2.

The value of t is determined from the condition on mean square error (MSE) falling below a particular value

$$Z(t) \leq \alpha \tag{17}$$

The method of obtaining the optimum values of the number of neurons in the hidden layer, and the Mutation and crossover fraction in the genetic algorithm parameters, is called parameter tuning, which is set by trials of different combination of the above parameters. The activation function used here is a sigmoid function given by:

$$f = \frac{1}{1 + e^{-a(x-c)}} \tag{18}$$

a and c has been considered as 1, and the bias in the network has been made 0.

Pseudo Code

Setting up the neural network

      Set iteration counter $t=0$;

      Set Initial X_n node = 0;

      Set Initial MSE=0.5;

Training of the network

While (MSE $''/> \alpha$)

    $t=t+1$;

Evaluate $Z(t)$ in terms of training set of inputs and nodes ;

Call GA function with $Z(t)$ as objective function

    Get **A** = solution given by GA ;

Evaluate **Y**(t) using A ;

Update node **X**$_n$=**Y**(t);

End

Value of **A** from last iteration = **A***

Testing of the network

Evaluate **Z***(t) in terms of testing set of inputs, nodes and **A***;

% **Z***(t) is the final MSE and **A*** is the required weights of the networks %

Function Genetic Algorithm
Put initial value (Z(t)):

Put Size of initial population;

Choose crossover operator and mutation operator;

Put mutation ratio (M);
Put crossover ratio (C);
Put generation size;

End

For every generation do:

    For every chromosome do:

        Encode the chromosome;

        If chromosome feasibility= positive;

            Evaluate **Z**(t);

        Else
        Back to encoding;
        End if

    Next:
    Compute chromosome ratio;
    Do

        Selection of specific gene population as obtained by chromosome ratio;
        Crossover;
        If chromosome feasibility= positive;

            Include it in initial population;

Else

Eliminate it;

End if

Iterate until crossover ratio (C) reached;

Chromosomes sorted in increasing order as per Z(t) value:

Select chromosomes keeping population size same as initial population size:

Eliminate the left ones:

Obtain chromosome ratio:

Do

Select 2 genes for mutation as per the chromosome ratio;

Mutate;

If chromosome feasibility= positive;

Include it in the initial population;

Else

Remove it from database;

End if

Iterate until mutation ratio (M) is reached;

Chromosomes sorted in increasing order as per Z(t) value:Select chromosomes keeping population size same as initial population size:

Eliminate the left ones:

While decided number of iterations reached or values within specified error limit;

5. Results

The matrices of distances required in all algorithms are defines by the Table 2 for the principal route, Table 3 for the capillary route:

	Barcelona	Zaragoza	Madrid	Cuenca	Teruel	Lleida
Barcelona	0	290.6	606.5	486.5	367.5	151.8
Zaragoza	290.6	0	320.9	255.8	176.0	141.1
Madrid	606.5	320.9	0	164.4	293.9	460.8
Cuenca	486.5	255.8	164.4	0	129.8	368.9
Teruel	367.5	176.0	293.9	129.8	0	249.9
Lleida	151.8	141.1	460.8	368.9	249.9	0

Table 3. Distance matrix. Madrid-Barcelona route

	Toledo	Bargas	Torrijos	Fuensalida	Recas	Illescas	Yuncos	Esquivias	Añover Tajo	Mocejón
Toledo	0	9.90	27.2	28.0	24.7	34.4	29.9	42.8	32.0	14.7
Bargas	9.90	0	24.4	24.5	15.1	26.5	22.0	34.8	26.9	10.6
Torrijos	27.22	24.4	0	10.8	37.8	50.4	45.6	58.9	51.4	35.1
Fuensalida	28.0	24.5	10.8	0	31.2	40.9	36.2	49.5	51.5	35.1
Recas	24.7	15.1	37.8	31.2	0	19.6	14.8	28.1	22.6	16.0
Illescas	34.4	26.5	50.4	40.9	19.6	0 0	4.80	8.50	19.5	25.6
Yuncos	29.9	22.0	45.6	36.2	8 8.14	4.80	0	13.4	15.0	21.1
Esquivias	42.8	34.8	58.9	49.5	28.1	8.50	13.4	0	21.9	34.0
Añover de Tajo	32.0	26.9	51.4	51.5	22.6	19.5	15.0	21.9	0	19.0
Mocejón	14.7	10.6	35.1	35.1	16.0	25.6	21.1	34.0	19.0	0

Table 4. Distance matrix. Capillary Routes

REFERENCE SOLUTION	Route 1	Route 2	Capillary Route	Total
Total distance (km)	1223	114	209.9	1558.1
Number of Cities visited	2	2	1	5
Time	14h41	1h28	2h41	18h50
Fuel Cost (€)	366.93	34.22	31.49	432.64

Table 5. Reference route provided by the company

RNNGA provides the following main route (see Figure 2):

Barcelona → Zaragoza → Madrid → Cuenca → Teruel → Lleida → Barcelona,

with a total distance of 1307.4 Km, only 84.4 km more than the reference route, but it presents better flexibility with two additional cities that are visited, employing 7 minutes more to cover the route than the reference route, with an extra cost of 20.19 €.

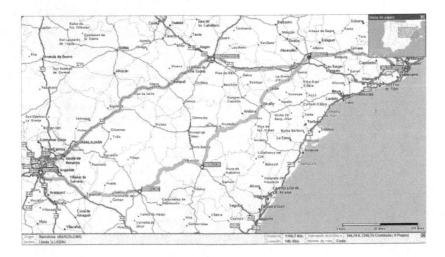

Figure 2. Optimal solution for the principal route obtained by RNNGA

The capillary route found by RNNGA is (see Figure 3):

Toledo → Torrijos → Bargas → Fuensalida → Recas → Añover de Tajo → Illescas → Yuncos → Esquivias → Mocejón → Toledo

with a distance of 216.3 Km, 6.4 km more than reference route, with a fuel cost of € 32.51, reducing 4 minutes the reference route.

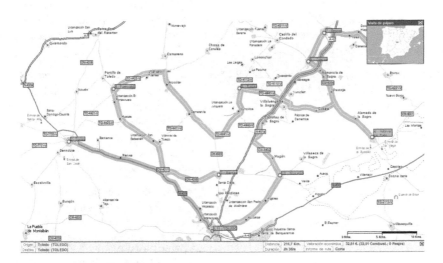

Figure 3. Route capillary provides by RNNGA

The total distance and the cost of the main routes will be added the Madrid-Toledo trajectory (57 Km), covered in 44 minutes with a cost of 17.11 € (fuel). Table 5 shows the results of the routes found by RNNGA.

	RNNGA			
	Route1	Route2	Capillary route	Total
Total distance (km)	1307.4	114	216.3	1637.7
Number of cities visited	5	2	1	8
Duration	14h48	1h28	2h38	18h54
Fuel Cost (€)	346,74	34.22	32.51	413.47

Table 6. Solution provided by the RNNGA.

6. Conclusions

A real case study has been considered, which the principal route consists in determining the route for sending a product set from Barcelona to Toledo (Spain). A first approach in this case study is to employ a route which passes through the maximum number of cities as possible minimizing costs, with the objective of maximise the flexibility. It will lead to the vehicle pick up or deliver products in those cities.

When the vehicle arrives from Madrid to Toledo (a direct route that will not be considered in the dual problem), the products require to be served in different towns close to Toledo. It is done following capillary routes. The case study considers a new vehicle that visits ten towns, starting and finishing in Toledo

The problem can be approximated to a series of problems similar to the vehicle routing problem with time windows (VRPTW). VRPTW tries to find a final solution including sub-path not connected and that meet the constraints of the travel salesman problem (TSP) considering the time windows constrains. The restrictions of Miller et al. (1960) have been used in order to reduce the computational cost. The main purposes of the method are to provide a quick solution and flexible enough to be used in a dynamic scheduling environment, and to develop a new solution procedure that is capable of exploiting the special characteristics of the problem.

This paper presents a meta-heuristic method that determines the routes that involve the major number of cities in order to increase flexibility, leading the vehicle deliver and pick up in these cities, trying to minimize the distance travelled and its costs.

A recurrent neural network approach is employed, which involves not just unsupervised learning to train neurons, but an integrated approach where Genetic Algorithm is utilized for training neurons so as to obtain a model with the least error.

Author details

Fausto Pedro García Márquez and Marta Ramos Martín Nieto

*Address all correspondence to: FaustoPedro.Garcia@uclm.es

Ingenium Research Group, Universidad Castilla-La Mancha, Ciudad Real, Spain

References

[1] Aleksander, I, & Morton, H. (1990). An introduction to neural computing, Chapman &Hall, London.

[2] DBK ((2009). Transporte de mercancías por carreteraDBK.

[3] Nilsson, N. J. Principles of artificial intelligence. New York: editing Birkhauser, (1982).

[4] Junger, M, Reinelt, G, & Rinaldi, G. (1997). The travelling salesman problem. In: Dell Amico M, Maffioli F and Martello S (eds) Annotated Bibliographies in. Combinatorial Optimization. John Wiley and Sons: Chichester, , 199-222.

[5] Blanco, A, Delgado, M, & Pegalajar, M. C. (2000). A genetic algorithm to obtain the optimal recurrent neural network. International Journal of Approximate Reasoning 23, , 67-83.

[6] Bourlard, H, & Wellekens, C. (1989). Speech pattern discrimination and multi-layered perceptrons. Computer Speech and Language 3, , 1-19.

[7] Le Cun YBoser B., Denker J., Henderson D., Howard, Hubbard, W. and Jackel L. ((1989). Backpropagation applied to handwritten zip code recognition. Neural Computation 1, , 541-551.

[8] Sejnowski, T, & Rosenberg, C. (1987). Parallel networks that learn to pronounce English text. Complex Systems 1, , 145-168.

[9] Waibel, A, Hanazawa, T, Hinton, G, Shikano, K, & Lang, K. (1989). Phoneme recognition using time delay neural networks. IEEE Transactions on Acoustics, Speech, and Signal Processing, 37, , 328-339.

[10] Bellmore, M, & Nemhauser, G. L. (1968). The traveling salesman problem: a survey, Operations Research, 16(3), , 538-588.

[11] Bodin, L. (1975). A toxonomic structure for vehicle routing and scheduling problems, Computers and Urban Society, 1, , 11-29.

[12] Gillett, B, & Miller, L. (1974). A heuristic algorithm for the vehicle dispatch problem, Operations Research, 22, , 340.

[13] Turner, W. C, Fouids, L, & Ghare, P. M. (1974). Transportation routing problem-a survey, AIIE Trans. 6, 4, , 288-301.

[14] Evans, S. R, & Norback, J. P. (1985). The impact of a decision-support system for vehicle routing in a foodservice supply situation, Journal of the Operational Research Society, , 36, 467-472.

[15] Faulin, J. (2003). Applying MIXALG procedure in a routing problem to optimize food product delivery, Omega, , 31, 387-395.

[16] Hsu, C. L, & Feng, S. (2003). Vehicle routing problem for distributing refrigerated food, Journal of the Eastern Asia Society for Transportation Studies, , 5, 2261-2272.

[17] Huey-Kuo ChenaChe-Fu Hsuehb, Mei-Shiang Changc ((2009). Production scheduling and vehicle routing with time windows for perishable food products, Computers & Operations Research, , 36, 2311-2319.

[18] Ioannou, G, Kritikos, M, & Prastacos, G. (2001). A greedy look ahead heuristic for the vehicle routing problem with time windows, Journal of the Operational Research Society, , 52, 523-537.

[19] Ma, H, Cheang, B, Lim, A, Zhan, L, & Zhu, Y. (2012). An investigation into the vehicle routing problem with time windows and link capacity constraints, Omega, , 40

[20] Prins ((2004). A simple and effective evolutionary algorithm for the vehicle routing problem, Computers & Operations Research, , 1985-2002.

[21] Prindezis, N, Kiranoudis, C. T, & Kouris, D. M. business fleet management service provider for central food market enterprises, Journal of Food Engineering, , 60, 203-210.

[22] Gendreau, M, Iori, M, Laporte, G, & Martello, S. (2006). A Tabu Search Algorithm for a Routing and Container Loading Problem. Transportation Science. 40(3), , 342-350.

[23] Tarantilis, C. D, & Kiranoudis, C. T. algorithm for the efficient distribution of perishable foods, Journal of Food Engineering, , 50, 1-9.

[24] Tarantilis, C. D, & Kiranoudis, C. T. (2002). Distribution of fresh meat. Journal of Food Engineering, , 51, 85-91.

[25] Lawler, E. L, Lenstra, J. K, Rinnooy, A. H. G, & Shmoys, D. B. eds.) ((1985). The Traveling Sales- man Problem. Wiley, New York.

[26] Gutin, G. and Punnen A (eds.) ((2002). The traveling salesman problem and its variations, Combinatorial Optimization.

[27] Ahr, D, & Reinelt, G. (2006). A tabu search algorithm for the minmax k-chinese postman problem. Computers & Operations Research, 33, , 3403-3422.

[28] Augerat, P, Belenguer, E, Benavent, J. M, Corbern, A, & Naddef, D. (1998). Separating capacity constraints in the CVRP using tabu search. European Journal of Operational Research, 106, , 546-557.

[29] Badeau, P, Gendreau, M, Guertin, F, Potvin, J-Y, & Taillard, E. (1997). A parallel tabu search heuristic for the vehicle routing problem with time windows. Transportation Research, 5, , 109-122.

[30] Barbarosoglu, G, & Ozgur, D. (1999). A tabu search algorithm for the vehicle routing problem. Computers and Operations Research, 26, , 255-270.

[31] Bianchessi, N, & Righini, G. (2007). Heuristic algorithms for the vehicle routing problem with simultaneous pick-up and delivery. Computers & Operations Research, 34, , 578-594.

[32] Brandao, J, & Mercer, A. (1997). A tabu search algorithm for the multi-trip vehicle routing and scheduling problem. European Journal of Operational Research, 100, , 180-191.

[33] Breedam, A. V. (2001). Comparing descent heuristics and metaheuristics for the vehicle routing problem. Computers & Operations Research, 24, , 289-315.

[34] Chao, I. M. (2002). A tabu search method for the truck and trailer routing problem. Computers & Operations Research, 29, , 33-51.

[35] Daniels, R. L, Rummel, J. L, & Schantz, R. (1998). A model for warehouse order picking. European Journal of Operational Research, 105, , 1-17.

[36] Garcia, B. L, Potvin, J. Y, & Rousseau, J. M. (1994). A parallel implementation of the tabu search heuristic for vehicle routing problems with time window constraints. Computers & Operations Research, 21, , 1025-1033.

[37] Gendreau, M, Laporte, G, & Sguin, R. (1996). A tabu search heuristic for the vehicle routing problem with stochastic demands and customers. Operations Research, 44, , 469-477.

[38] Gendreau, M, Laporte, G, & Vigo, D. (1999). Heuristics for the traveling salesman problem with pickup and delivery. Computers & Operations Research, 26, , 699-714.

[39] Hertz, A, Laporte, G, & Mittaz, M. (2000). A tabu search heuristic for the capacitated arc routing problem. Operations Research, 48, , 129-135.

[40] Lau, H. C, Sim, M, & Teo, K. M. (2003). Vehicle routing problem with time windows and a limited number of vehicles. European Journal of Operational Research, 148, , 559-569.

[41] Malek, M, Guruswamy, M, Pandya, M, & Owens, H. (1989). Serial and parallel simulated annealing and tabu search algorithms for the traveling salesman problem. Annals of Operations Research, 21, , 59-84.

[42] Montane, F. A. T, & Galvao, R. D. (2006). A tabu search algorithm for the vehicle routing problem with simultaneous pick-up and delivery service. Computers & Operations Research, 33, , 595-619.

[43] Nanry, W. P, & Barnes, J. W. (2000). Solving the pickup and delivery problem with time windows using reactive tabu search. Transportation Research Part B: Methodological, 34, , 107-121.

[44] Osman, I. H. (1993). Metastrategy simulated annealing and tabu search algorithms for the vehicle routing problem. Annals of Operations Research, 41, , 421-451.

[45] Potvin, J. Y, Kervahut, T, Garcia, B. L, & Rousseau, J. M. (1996). The vehicle routing problem with time windows- part I: Tabu search. INFORMS Journal of Computing, 8, , 158-164.

[46] Scheuerer, S. (2006). A tabu search heuristic for the truck and trailer routing problem. Computers & Operations Research, 33, , 894-909.

[47] Semet, F, & Taillard, E. (1993). Solving real-life vehicle routing problems efficiently using taboo search. Annals of Operations Research, 41, , 469-488.

[48] Tarantilis, C. D. (2005). Solving the vehicle routing problem with adaptive memory programming methodology. Computers & Operations Research, 32, , 2309-2327.

[49] Tarantilis, C. D, & Kiranoudis, C. T. (2007). A flexible adaptive memory-based algorithm for real-life transportation operations: Two case studies from dairy and construction sector. European Journal of Operational Research, 179, , 806-822.

[50] Carpaneto, G, & Toth, P. (1980). Some new branching and bounding criteria for the asymmetric traveling salesman problem, Management Science 26, , 736-743.

[51] Gen, M, & Cheng, R. Genetic Algorithms and Engineering Design, Wiley, New York, (1997).

[52] Fischetti, M, & Toth, P. (1989). An additive bounding procedure for combinatorial optimization problems, Operations Research 37, , 319-328.

[53] Finke, G, Claus, A, & Gunn, E. network flow approach to the traveling salesman problem, Congressus Numerantium, 41, , 167-178.

[54] Glover, F. (1990). Artificial intelligence, heuristic frameworks and tabu search, Managerial & Decision Economics, 11, , 365-378.

[55] Gouveia, L, & Pires, J. M. (1999). The asymmetric travelling salesman problem and a reformulation of the Miller-Tucker- Zemlin constraints, European Journal of Operational Research, 112, , 134-146.

[56] Kirkpatrick, S, & Gelatt, C. D. Jr. and Vecchi M.P., ((1985). Configuration space analysis of travelling salesman problem, Journal Physique, 46, , 1277-1292.

[57] Lin, S, & Kernighan, B. W. (1973). An effective heuristic algorithm for travelling salesman problem, Operations Research, , 498-516.

[58] Lysgaard, J. (1999). Cluster based branching for the asymmetric travelling salesman problem, European Journal of Operational Research, 119, , 314-325.

[59] Potvin, J. Y. (1996). Genetic algorithms for the travelling salesman problem, Annals of Operations Research, 63, , 339-370.

[60] Shirrish, B, Nigel, J, & Kabuka, M. R. (1993). A boolean neural network approach for the travelling salesman problem, IEEE Transactions on Computers, 42, , 1271-1278.

[61] Wong, R. (1980). Integer programming formulations of the travelling salesman problem, in: Proceedings of the IEEE International Conference of Circuits and Computers, , 149-152.

[62] Moon, C, Kim, J, Choi, G, & Seo, Y. (2002). An efficient genetic algorithm for the travelling salesman problem with precedence constraints. European Journal of Operational Research, 140, , 606-617.

[63] Little, J, Murty, K, Sweeney, D, & Karel, C. (1963). An algorithm for the traveling salesman problem, Operations Research 11, , 972-989.

[64] Balas, E, & Toth, P. (1985). Branch and bound method s. In: Lawler EL, Lenstra JK, Rinnooy Kan AHG, Shmoys DB, editors. The traveling salesman problem. New York: Wiley, 1985. , 361-402.

[65] Verhoeven, M. G. A, Aarts, E. H. L, & Swinkels, P. C. J. opt algorithm for the travelling salesman problem, Future Generation Computer Systems 11, , 175-182.

[66] Burke, L. I. (1994). Neural methods for the traveling salesman problem: insights from operations research, Neural Networks, 7(4), , 681-690.

[67] Miller, C. E, Tucker, A. W, & Zemlin, R. A. (1960). Integer programming formulation of travelling salesman problems, J. ACM, 3, , 326-329.

Chapter 2

Modeling Engineering Management Decisions with Game Theory

William P. Fox

Additional information is available at the end of the chapter

1. Introduction

Engineering management is a management process that blends of engineering and business processes. As such the ability to make decision as well as model the decision making process may be critical steps in the process. In game theory, we employ the process to gain insights into possible courses of action from each player assuming the players are rational, that is they want to maximize their gains.

In many business situations, two or more decision makers simultaneously and without communications choose courses of actions, and the action chosen by each affects the payoff or gains earned by all the other players. For example, consider a fast food chain such as Burger King. If they choose an advertising strategy with pricing not only do they help their payoffs but their choices affect all other fast food chains. Each company's decision affect the revenues, profits, of loses of the other fast food chains.

Game theory is useful in analyzing decisions in cases where two or more decision makers have conflicting interest. Most of what we present here concerns only the two person game but we will also briefly examine the n-person game.

In two person games, each of the players has strategies or courses of action that they might choose. These courses of action lead to outcomes or payoffs to the decision maker and these payoffs might be any values (positive, negative, or zero). These payoffs are usually presented in a payoff matrix such as the general one presented in Table 1. In table 1 player 1, whom we will call Rose, might have m course of actions available and player 2, whom we will call Colin, may have n courses of actions available. These payoffs values might have come from ordinal utilities or cardinal utilities. For more information about obtaining payoff values please see the additional reading (Straffin, 2004; Von Neumann & Morgenstern, 2004).

Game theory is the branch of mathematics and decision theory concerned with strategic decisions when two or more players compete. The problems of interest involve multiple participants, each of whom has individual strategies related to a common system or shared resources. Because game theory arose from the analysis of competitive scenarios, the problems are called *games* and the participants are called *players*. But these techniques apply to more than just sport, and are not even limited to competitive situations. In short, game theory deals with any problem in which each player's strategy depends on what the other players do. Situations involving interdependent decisions arise frequently, in all walks of life. A few examples in which game theory might be used include:

• Friends choosing where to go have dinner

• Couples deciding between going to ballet or a sporting event

• Parents trying to get children to behave

• Commuters deciding how best to travel to work

• Businesses competing in a fair market

• Diplomats negotiating a treaty

• Gamblers betting in a game of chance

• Military strategists weighing alternatives, such as attack or defend

• Governmental diplomacy options for sanctions or actions

• Pitcher-batter duel in baseball or penalty kicker-goalie duel in soccer

All of these situations call for strategic thinking, making use of available information to devise the best plan to achieve one's objectives. Perhaps you are already familiar with assessing costs and benefits in order to make informed decisions between several options. Game theory simply extends this concept to interdependent decisions, in which the options being evaluated are functions of the players' choices or their utility.

Player 1, Rose's Strategies			Player 2, Colin's Strategies	
	Column 1	Column 2	...	Column n
Row 1	$M_{1,1}N_{1,1}$	$M_{1,2}N_{1,2}$	...	$M_{1,n}N_{1,n}$
Row 2	$M_{2,1}N_{2,1}$	$M_{2,2}N_{2,2}$	...	$M_{2,n}N_{2,n}$
.	.	.	.	.
.	.	.	.	.
.	.	.	.	.
Row m	$M_{m,1}N_{m,1}$	$M_{m,2}N_{m,2}$	...	$M_{m,n}N_{m,n}$

Table 1. Payoff Matrix, M, of a two-person total conflict game

Consider the situation where two major discount stores want to come into the same region. We will call these two major discount stores, Target and Wal-Mart. Each store can decide whether to build their store in the region's larger city or in the region's smaller city. The stores desire the bigger market share of the consumers that yields more profits for their respective company. Experts have estimated the market share in the region for the larger and smaller city building options based upon 100% of the consumer market and income of the region. Based upon this market research, table 1, what decisions should each store make? As we will show later in this chapter, the best decision for each store is to locate in the larger city.

		Wal-Mart	
		Large City	Small City
Target	Large City	(60,40)	(75,25)
	Small City	(50,50)	(58,42)

Table 2. Target versus Wal-Mart

Two types of games will be presented in this chapter: total conflict games and partial conflict games. Game theory then is the study of decisions where the outcome to the decision maker depends not only on what he does, but the decision of one or more additional players. We classify the games depending upon whether the conflict between the players is *total* or *partial*. A total conflict game is a game where the sum of values in each cell of the payoff matrix, $M_{ij}+N_{ij}$ either always equals 0 or always equals the same constant for each ij pair. In a partial conflict game, this sum does not always equal 0 or the same constant. We begin our discussion with the total conflict game described in Table 1.

2. Two–person total conflict games

We begin with characteristics of the two-person total conflict game:

1. There are two persons (called the row player who we will refer to as Rose and the column player who we will refer to as Colin).

2. Rose must choose 1 of m strategies and Colin must choose one of n strategies.

3. If Rose chooses the ith strategy and Colin the jth strategy then Rose receives a payoff of a_{ij} and Colin loses an amount a_{ij}.

4. There are two types of possible solutions. Pure strategy solutions are when each player achieves their best outcomes by always choosing the same strategy in repeated games. Mixed strategy solutions are when players play a random selection of their strategies in order to obtain their best outcomes in repeated games.

Games might be presented either in decision tree or payoff format. In a decision tree for sequential games, we look ahead and reason back. In simultaneous games, we use payoff matrices as shown in Table 1. This is a total conflict game if and only $M_{i,j}+N_{i,j}$ equals either 0 or the same constant for all i and j.

For example, if a player wins x when the other player loses x then their sum is zero or in business marketing strategy based upon 100% if one player get $x\%$ of the market then the other player gets $y\%$ such that their sum is $x\%+y\%=100$. Given a simple payoff matrix we look for the Nash equilibrium as the solution first with movement diagrams.

Example 1 *Target* versus *Wal-Mart*

Suppose Large City is located near Small City. Now assume a smaller grocery store such as Target will locate a franchise in either Large City or Small City. Further, a "mega grocery store" franchise such as Wal-Mart is making the same decision – they will locate either in Large City or Small City. Analysts have estimated the market shares and we place both sets of payoffs in a single game matrix. Listing the row player's payoffs first, we have the payoff as shown in Table 2. We apply the movement diagram, were we draw arrows in each row (vertical arrow) and column (horizontal arrow) from the smaller payoff to the larger payoff.

		Wal-Mart	
	Large City		Small City
Target — Large City	60, 40		75, 25
Target — Small City	50, 50		58, 42

Table 3. Payoff matrix for example 1

Note all arrows point into the payoff (60,40) at (Large City, Large City) strategies for both players and no arrow exits that outcome, see Table 3. This indicates that neither player can unilaterally improve their solution. This stable situation is called a Nash Equilibrium. Often payoff matrices and movement diagrams may get convoluted or the arrows do not point to one or more points. In those more complex two-person games we offer linear programming as the solution method.

3. Linear programming of total conflict games

Every total conflict game may be formulated as a linear programming problem. Consider a total conflict two person game in which maximizing player X has m strategies and minimizing player Y has n strategies. The entry (M_{ij}, N_{ij}) from the ith row and jth column of the payoff matrix represents the payoff for those strategies. We present the following formulation using the elements of M for the maximizing a player that provides results for the value of the game and the probabilities x_i (Fox, 2010; Fox, 2012; Winston, 2003). We note that if there are negative values in the payoff matrix then we need a slight modification to the formulation. We suggest the method by Winston (2003) to replace any variable that could take on negative values with the difference in two positive variables, $V_j - V'_j$. We only assume that the value of the game could be positive or negative. The other values we are looking for are probabilities that are always non-negative.

Maximize V

Subject to

$$
\begin{aligned}
N_{1,1}x_1 + N_{2,1}x_2 + ... + N_{m,1}x_n - V &\geq 0 \\
N_{2,1}x_1 + N_{2,2}x_2 + ... + N_{m,2}x_n - V &\geq 0 \\
&... \\
N_{m,1}x_1 + N_{m,2}x_2 + ... + N_{m,n}x_n - V &\geq 0 \\
x_1 + x_2 + ... + x_n &= 1 \\
Nonnegativity&
\end{aligned}
\tag{1}
$$

where the weights x_i yields Rose's strategy and the value of V is the value of the game to Colin.

Maximize v

Subject to

$$
\begin{aligned}
M_{1,1}y_1 + M_{2,1}y_2 + ... + M_{m,1}y_n - v &\geq 0 \\
M_{2,1}y_1 + M_{2,2}y_2 + ... + M_{m,2}y_n - v &\geq 0 \\
&... \\
M_{m,1}y_1 + M_{m,2}y_2 + ... + M_{m,n}y_n - v &\geq 0 \\
y_1 + y_2 + ... + y_n &= 1 \\
Nonnegativity&
\end{aligned}
\tag{2}
$$

where the weights y_i yield Colin's strategy and the value of v is the value of the game to Rose.

Our two formulations for this problem are for Rose and Colin, respectively:

Maximize V_c

Subject to:

$40 \, x_1 + 50x_2 - V_c \geq 0$

$25x_1 + 42x_2 - V_c \geq 0$

$x_1 + x_2 = 1$

$x_1, x_2, V_c \geq 0$

Maximize V_r

Subject to:

$$60\,y_1 + 75y_2 - V_r \geq 0$$

$$50y_1 + 58y_2 - V_r \geq 0$$

$$y_1 + y_2 = 1$$

$$y_1,\,y_2,\,V_r \geq 0$$

If we put our example into our two formulations and solve, we get the solution $y_1=1, y_2=0$ and $Vr=60$ from formulation (1) and $x_1=1, x_2=0$, and $Vc=40$ from formulation (2). The overall solution is (Large City, Large City) with value (60,40).

4. Constant–sum to zero–sum

The primal-dual only works in the zero-sum game format. We may convert this game to the zero-sum game format to obtain. Since this is a constant sum game, all outcomes sum to 100. This can be converted to a zero sum game through the positive linear function, $y=x-20$. Use any two pairs of points and obtain the equation of the line and then make the slope positive. Using this transformation $x_1=x-20$ we can obtain the payoffs for the row player in the zero-sum game. The new zero-sum payoff matrix may be written as follows:

		Target	
		Large City	Small City
Wal-Mart	Large City	40 ⟵	55 ↑
	Small City	30 ↑ ⟵	38

Table 4. Add Caption

For a zero-sum game we can again look at movement diagrams, dominance, or linear programming. If one Rose's information is present representing the zero sum game then only assume Colin's values are the negative of Rose's. We apply the movement diagram as before and place the arrows accordingly. The arrows point in and never leave 40. The large city, large city strategy is the stable pure strategy solution. We define dominance as:

A strategy A dominates a strategy B if every outcome in A is at least as good as the corresponding outcome in B, and at least one outcome in A is strictly better than the corresponding outcome in B. Dominance Principle: A rational player should never play a dominated strategy in a total conflict game.

In this case the small city strategy payoffs for Rose are dominated by the large city strategy payoffs, thus we would never play small city. For Colin, the large city is better than the small city, so the large city dominates. Since large city, large city is the dominate strategy the solution is (40,-40).

If we use linear programming we only need a single formulation of the linear program. The row player maximizes and the column player minimizes with rows' values. This constitutes a primal and dual relationship. The linear program used for Rose in the zero-sum games is:

Maximize V

Subject to:

$$a_{1,1}x_1 + a_{1,2}x_2 + ... + a_{1,n}x_n - V \geq 0$$
$$a_{2,1}x_1 + a_{2,2}x_2 + ... + a_{2,n}x_n - V \geq 0$$
$$.$$
$$.$$
$$.$$
$$a_{m,1}x_1 + a_{m,2}x_2 + ... + a_{m,n}x_n - V \geq 0$$
$$x_1 + x_2 + ... + x_n = 1$$
$$V, x_i \geq 0$$

(3)

where V is the value of the game, $a_{m,n}$ are payoff-matrix entries, and x's are the weights (probabilities to play the strategies). We place these payoffs into our formulation:

Max V_r

Subject to:

$40x_1 + 30x_2 - V_r \geq 0$

$55x_1 + 38x_2 - V_r \geq 0$

$x_1 + x_2 = 1$

$x_1, x_2, V_r \geq 0$

The optimal solution strategies found are identical as before with both players choosing Large City as their best strategy. This indicates that neither player can unilaterally improve, a stable situation that we refer to as a Nash Equilibrium.

A Nash equilibrium is an outcome where neither player can benefit by departing unilaterally from its strategy associated with that outcome.

We conclude our discussion of total conflict games with the analysis that linear programming may always be used all total conflict games but is most suitable for large games between two players each having many strategies (Fox, 2010; Fox 2012).

5. Two–person partial conflict games

In the previous example, the conflict between the decision makers was total in the sense that neither player could improve without hurting the other player. If that is not the case, we classify the game as *partial conflict* as illustrated in the next example. Assume we have new market share analysis for our two stores as shown in Table 3 where the sums are not all equal to same constant.

	Wal-Mart		
	Large City		**Small City**
Large City	65, 25	$\Longrightarrow$	50, 45
Target	$\Uparrow$		$\Downarrow$
Small City	55, 40	$\Longleftarrow$	62, 28

Table 5. Revised market shares

We begin with the movement diagram where the arrows do not find a stable point. In those cases, we need to find the equalizing strategies to find the Nash equilibrium. Other solution methods are found in the additional reading, we will describe only the use of linear programming as the method to use when the movement diagram fails to yield a stable point. Additionally, Gillman and Housman (2009) state that every partial conflict game also has equalizing strategy equilibrium even if it also has a pure strategy equilibrium.

Since both players are maximizing their payoffs we use the linear programming formulation presented as equation (1) and (2).

This yields two separate linear programs.

Maximize V

Subject to:

$65\, y_1 + 50\, y_2 - V \geq 0$

$55\, y_1 + 62\, y_2 - V \geq 0$

$y_1 + y_2 = 1$

$y_1,\, y_2,\, V \geq 0$

The solution is $y_1 = 6/11$, $y_2 = 5/11$ and $V = 58.182$

The other second LP formulation is

Maximize v

Subject to:

$25\, x_1 + 40\, x_2 - v \geq 0$

$45\, x_1 + 28\, x_2 - v \geq 0$

$x_1 + x_2 = 1$

$x_1,\, x_2,\, v \geq 0$

The solution is $x_1=17/32$, $x_2=15/32$ and $v=34.375$.

The results state that the two stores must play their two strategies each a proportion of the time that they compete in order to obtain their best outcomes.

Another option available in partial conflict games is to consider allowing cooperation and communications between the game's players. This allows for first moves, threats, promises, and combinations of threats and promises in order to obtain better outcomes. We call this strategic moves (Straffin, 2004).

6. Moving first or committing to move first

We now assume both players can communicate their plans or their moves to the second player. If Target can move first they can choose Large City or Small City. Examining the movement diagram, they should expect Wal-Mart's responses as follows:

If Target plays Large City, Wal-Mart plays Small City resulting in the outcome (50,45).

If Target plays Small City, Wal-Mart plays Large City plays resulting in the outcome (55,40).

Target prefers 55 to 50 so they would play Small City. If Target forces Wal-Mart to move first then the choices are between (65,25) and (62,28) of which Wal-Mart prefers (62,28). Having Wal-Mart to move first gets Target a better outcome. How to get this to occur as well as the credibility of the first move is a concern.

7. Issuing a Threat

In general we describe the concept of issuing a threat. Rose may have a threat to deter Colin from playing a particular strategy. A threat must satisfy three conditions:

Conditions for a Threat by Rose

1. Rose communicates that she will play a certain strategy contingent upon a previous action of Colin.

2. Rose's action is harmful to Rose.

3. Rose's action is harmful to Colin.

In our game example there is no valid threat.

We present the classic game of chicken to show a valid threat.

In the game of Chicken, Rose wants Colin to play Swerve. Therefore she makes the threat on Colin's Not Swerve to deter him from choosing that strategy. Examining the movement diagram,

	Colin	
	Swerve	Not Swerve
Swerve	(3, 3)	(2,4)
Rose		
Not Swerve	(4, 2)	(1, 1)

Table 6. Chicken Game

Normally, if Colin plays Not Swerve, Rose plays Swerve yielding (2, 4)

In order to harm herself, Rose must play Not Swerve. Thus the potential threat must take the form:

If Colin plays Not Swerve, then Rose plays Not Swerve yielding (1, 1).

Is it a threat? It is contingent upon Colin choosing Not Swerve. Comparing (2, 4) and (1, 1), we see that the threat is harmful to Rose and is harmful to Colin. It is a threat and effectively eliminates the outcome (2, 4) making the game,

	Colin	
	Swerve	Not Serve
Swerve	(3,3) ⟹	(2,4)
Rose	⇓	⇑
Not Swerve	(4,2) ⟸	(1,1)

Table 7. Chicken game with a threat.

Colin still has a choice of choosing Swerve or Not Swerve. Using the movement diagram, he analyzes his choices as follows:

If Colin selects Swerve, Rose chooses Not Swerve yielding (4, 2).

If Colin chooses Not Swerve, Rose chooses Not Swerve yielding (1, 1) (because of Rose's threat).

Thus Colin's choice is between a payoff of 2 and 1. He should choose Swerve yielding (4, 2). If Rose can make her threat credible, she can secure her best outcome.

8. Issuing a promise

In our Target versus Wal-Mart game there is no promise, so again we illustrate with the classic game of chicken. Again, if Colin has the opportunity to move first or is committed to (or possibly considering) Not Swerve, Rose may have a promise to encourage Colin to play Swerve instead. A promise must satisfy three conditions:

Conditions for a Promise by Rose

1. Rose communicates that she will play a certain strategy contingent upon a previous action of Colin.

2. Rose's action is harmful to Rose.

3. Rose's action is beneficial to Colin.

In the game of Chicken, Rose wants Colin to play Swerve. Therefore she makes the promise on Colin Swerve to sweeten the pot so he will choose Swerve. Examining the movement diagram,

Normally, if Colin plays Swerve, Rose plays Not Swerve yielding (4, 2).

In order to harm herself, she must play Swerve. Thus the promise takes the form:

If Colin plays Swerve, then Rose plays Swerve yielding (3, 3).

Is it a promise? It is contingent upon Colin choosing Swerve. Comparing the normal (4, 2) with the promised (3, 3), we see that the promise is harmful to Rose and is beneficial to Colin. It is a promise and effectively eliminates the outcome (4, 2) making the game,

		Colin		
		Swerve		Not Swerve
	Swerve	(3, 3)	⟹	(2, 4)
Rose				↑
	Not Swerve	Eliminated by Promise		(1, 1)

Table 8. Chicken game with a promise.

Colin still has a choice of choosing Swerve or Not Swerve. Using the movement diagram, he analyzes his choices as follows:

If Colin selects Swerve, Rose chooses Swerve yielding (3, 3) as promised.

If Colin chooses Not Swerve, Rose chooses Swerve yielding (2, 4).

Thus Colin's choice is between payoffs of 3 and 4. He should choose Not Swerve yielding (2, 4). Rose does have a promise. But her goal is for Colin to choose Swerve. Even with the promise eliminating an outcome, Colin chooses Not Swerve. The promise does not work. If both make a promise then perhaps (3, 3) is the outcome.

In summary, the game of Chicken offers many options. If the players choose conservatively without communication, the maximin strategies yields (3, 3), which is unstable: both players unilaterally can improve their outcomes. If either player moves first or commits to move first they can obtain their best outcome. For example, Rose can obtain (4, 2) which is a Nash equilibrium. If Rose issues a threat, she can eliminate (2, 4) and obtain (4, 2). A promise by Rose eliminates (4, 2) but results in (2, 4) which does not improve the (3, 3) likely outcome without communication.

8. 1. A combination threat and promise

Consider the following game:

		Colin		
		C1		C2
	R1	(2, 4)	⟵	(3, 3)
Rose		↑		↓
	R2	(1, 2)	⟵	(4, 1)

Table 9. Rose versus Colin payoffs.

The movement diagram shows that (2,4). is the Nash equilibrium. Without communication, Colin gets his best outcome, but can Rose do better that (2, 4) with a strategic move?

Rose First If Rose moves R1, Colin should respond with C1 yielding (2, 4). If Rose moves R2, Colin responds with C1 yielding (1, 2). Rose's best choice is (2, 4), no better than the likely conservative outcome without communication.

Rose Threat Rose wants Colin to play C2. Normally, if Colin plays C1, Rose plays R1 yielding (2, 4). To hurt herself she must play R2 yielding (1, 2). Comparing the normal (2, 4) and (1, 2), the threat is contingent upon Colin playing C1, hurts Rose and hurts Colin. It is a threat and effectively eliminates (2, 4) yielding

		Colin		
		C1		C2
	R1	eliminated		(3, 3)
Rose				↓
	R2	(1, 2)	⟵	(4, 1)

Table 10. Result of threat alone by Rose.

Does the threat deter Colin from playing C1? Examining the movement diagram, If Colin plays C1 the outcome is (1, 2). If Colin plays C2 the outcome is (4, 1). Colin's best choice is still C1. Thus there is a threat, but it does not work. Does Rose have a promise that works by itself?

Rose Promise Rose wants Colin to play C2. Normally, if Colin plays C2, Rose plays R2 yielding (4, 1). To hurt herself she must play R1 yielding (3, 3). Comparing the normal (4, 1) with the promised (3, 3), the move is contingent upon Colin playing C2, hurts Rose and is beneficial to Colin. It is a promise and effectively eliminates (4, 1) yielding

		Colin	
		C1	C2
	R1	(2, 4) ⟸	(3, 3)
Rose		↑	
	R2	(1, 2)	eliminated

Table 11. Result from promise alone by Rose.

Does the promise motivate Colin to play C2? Examining the movement diagram, if Colin plays C1 the outcome is (2, 4). If Colin plays C2 the outcome is (3, 3). Colin's best choice is still C1 for (2, 4). Thus there is a promise, but it does not work. What about combining both the threat and the promise?

Combination Threat and Promise We see that Rose does have a threat that eliminates an outcome but does not work by itself. She also has a promise that eliminates an outcome but does not work by itself. In such situations, we can examine issuing both the threat and the promise to eliminate two outcomes to determine if a better outcome results. Rose's threat eliminates (2, 4), and Rose's promise eliminates (4, 1). If she issues both the threat and the promise, the following outcomes are available.

		Colin	
		C1	C2
	R1	eliminated	(3, 3)
Rose			
	R2	(1, 2)	eliminated

Table 12. Result of combination of threat and promise.

If Colin plays C1 the result is (1, 2), and choosing C2 yields (3, 3). He should choose C2, and (3, 3) represents an improvement for Rose over the likely outcome without communication (2, 4).

Credibility Of course, commitments to first moves, threats and promises must be made credible. If Rose issues a threat, and Colin chooses to Not Swerve anyway, will Rose will carry out her threat and crash (1, 1) even though that action no longer promises to get her the outcome (4, 2)? If Colin believes that she will not carry through on her threat, he will ignore the threat. In the game of Chicken, if Rose and Colin both promise to Swerve and Colin believes Rose's promise and executes Swerve, will Rose carry out her promise to Swerve and accept (3, 3) even though (4, 2) is still available to her? One method for Rose to gain credibility is to lower one or more of her payoffs so that it is obvious to Colin that she will execute the stated move. Or, if possible, she may make a *side payment* to Colin to increase his selected payoffs in order to entice him to a strategy that is favorable to her and is now favorable to him because of the side payment. These ideas are pursued further in the exercises.

An inventory of the strategic moves available to each player is an important part of determining how a player should act. Each player wants to know what strategic moves are available to each of them. For example if Rose has a first move and Colin has a threat, Rose will want to execute her first move before Colin issues his threat. The analysis requires knowing the rank order of the possible outcomes for both players. Once a player has decided which strategy he wants the opposing player to execute, he can then determine how the player will react to any of his moves.

As alluded earlier maybe the better option is to go to arbitration. We discuss that next.

9. Nash arbitration

In the bargaining problem, Nash (1950) developed a scheme for producing a single fair outcome. The goals for the Nash arbitrations scheme are that the result will be at or above the status quo point for each player and that the result must be "fair".

Nash introduced the following terminology:

Status Quo Point (We will typically use the intersection of Rose's Security Level and Colin's Security Level; the Threat positions may also be used).

Negotiation Set: Those points in the Pareto Optimal Set that are at or above the "Status Quo" of both players.

We use Nash's four axioms that he believed that a reasonable arbitration scheme should satisfy rationality, linear invariance, symmetry, and invariance. A good discussion of these axioms and can be found in Straffin (2004, p.104-105). Simply put the Nash Arbitration point is the point that follows all four axioms. This leads to Nash's Theorem stated below:

Nash's Theorem: There is one and only one arbitration scheme which satisfies Axioms 1 through 4. It is this: if the *status quo* $SQ = (x_0, y_0)$, then the arbitrated solution point N is the point (x, y) in the polygon with $x \geq x_0$ and $y \geq y_0$ which *maximizes the product:* $(x - x_0)(y - y_0)$.

Let's examine this geometrically first as it will provide insights into using calculus methods. We produce the contour plot of our nonlinear function: $(x - x_0)(y - y_0)$ when our status quo point is assumed to be *(0,0)*. It is obvious that the northeast (NE) corner of quadrant 1 is where this function is maximized. This is illustrated in figure 1 below.

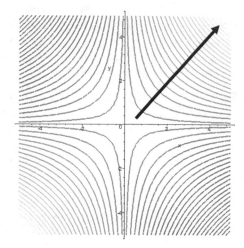

Figure 1. Contour plot for $(x*y)$. We note that the direction of maximum increase is NE as indicated by the arrow.

We need a few more definitions to use this Nash Arbitration.

In his theory for the arbitration and cooperative solutions, Nash (1950) stated the "reasonable" solution should be Pareto optimal and will be at or above the security level. The set of outcomes that satisfy these two conditions is called the *negotiation set*. The line segments that join the negotiation set must form: a convex region as shown in Nash's proof. Methodologies for solving for this point use basic calculus, algebra, and geometry.

For any game theory problem, we next overlay the convex polygon onto our contour plot. The most NE point in the feasible region is our optimal point and the Nash Arbitration point. This will be where the feasible region is tangent to the hyperbola. It will always be on the line segment that joins the negotiation set. This is simply a constrained optimization problem. We can convert to a single variable problem as we will illustrate later in our example.

In our example, we will use the security value as the status quo point to use in the Nash arbitration procedure. We additionally define the procedure to find the security value as follows:

In a non-zero-sum game, Rose's optimal strategy in Rose's game is called Rose's **Prudential Strategy**, the value is called Rose's **Security level**. Colin's optimal strategy in Colin's game is called Colin's **Security level**. We will illustrate this during the solution to find the Nash arbitration point in the following example.

To find the security level (status quo point) we look at the following two separate games extracted from the original game and use movement diagrams, dominance, or our linear programming method to solve each game for those players' values.

		Colin	
		C	D
Rose	A	(2,6)	(10,5)
	B	(4,8)	(0,0)

Table 13. Finding the *security levels* in a non-zero sum game

In a prudential strategy, we allow a player to find their optimal strategy in their own game. For Rose, she would need to find her optimal solution in her own game. Rose's game below has a mixed strategy solution; V=10/3.

		Colin	
		C	D
Rose	A	2	10
	B	4	0

Table 14. Rose's game for finding her Prudential strategy.

For Colin, he would need to find his optimal solution in his own game. Colin's game below has a pure strategy solution, V=6.

		Colin	
		C	D
Rose	A	6	5
	B	8	0

Table 15. Colin's game for finding his Prudential strategy.

The status quo point or security level from the Prudential strategy is found to be *(10/3, 6)*. We will use this point in the formulation of the Nash arbitration.

10. Finding the Nash arbitration point

We use the nonlinear programming method described by Fox (2010,2012). We set up the convex polygon (constraints) for the function that we want to maximize, which is $(x - \frac{10}{3}) \cdot (y - 6)$. The convex polygon is the convex set from the values in the pay-off matrix. Its boundary and interior points represent all possible combinations of strategies. Corner points represent **pure** strategies. All other points are mixed strategies. Occasionally, a pure strategy is an interior point. Thus, we start by plotting the strategies from our payoff matrix set of values { (2,6), (4,8), (10,5), (0,0)}, see figure 2.

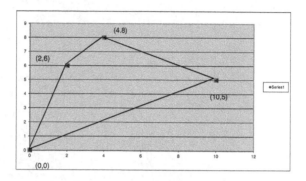

Figure 2. Payoff Polygon,

We note that our convex region has four sides whose coordinates are our pure strategies. We use the point-slope formula to find the equations of the line and then test points to transform the equations to inequalities. For example, the line form *(4,8)* to *(10,5)* is $y =$

-.5 x + 10. We rewrite as $y + .5x = 10$. Our test point *(0, 0)* shows that our inequality is

$0.5x + y \leq 10$. We use this technique to find all boundary lines as well as add our security levels as lines that we need to be above.

The convex polygon is bounded by the following in equalities:

$-.5x + y \leq 10$
$-3x + y \leq 0$
$0.5x - y \leq 0$
$-x + y \leq 4$
$x \geq x^*$
$y \geq y^*$

where x^* and y^* are the security levels (10/3,6).

The NLP formulation (Winston, 2003; Fox, 2012) to find the Nash arbitration value following the format of equation is as follows:

Maximize

$$(x - \frac{10}{3}) \cdot (y - 6) \tag{4}$$

Subject to:

$0.5x + y \leq 10$

$-3x + y \leq 0$

$0.5x - y \leq 0$

$-x + y \leq 4$

$x \geq \dfrac{10}{3}$

$y \geq 6$

We display the feasible region graphically in Figure 3 The feasible region is the solid region. From the figure we can approximate the solution as the point of tangency between the feasible region and the hyperbolic contours in the NE region.

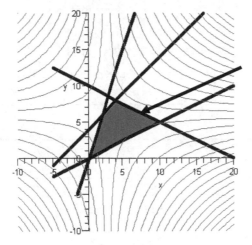

Optimal solution is the point of tangency.

Figure 3. Convex polygon and function contour plot.

Since we visually see that the solution must fall along the line segment y=-0.5 x+ 10. We may use simple calculus.

Maximize $(x - \dfrac{10}{3}) \cdot (y - 6)$

Subject to: $y = -.5x + 10$

We substitute to obtain a function of one variable,

Maximize $(x - 10/3)((-.5x + 10 - 6)$ or

Maximize $5x^2 + 34/6\, x - 40/3$

We find $\frac{df}{dx} = 0 = -x + 34/6$.

We find $x=17/3$

The second derivative test confirms we found a maximum.

We substitute $x= 17/3$ back into $y=-.5x+10$ to obtain $y= 43/6$. This point (17/3, 43/6) is the Nash Arbitration point. Our optimal solution, the Nash arbitration point is found to be $x =5.667$ and $y = 7.167$ and the value of the objective function payoff of 2.72.

How do we obtain this value in a particle manner? An arbitrator plays the strategies BC (4,8) and AD (10,5) as follows described below.

We can solve two equations and two unknowns from our strategies BC and AD equal to our Nash arbitration point.

$$\begin{bmatrix} 4 & 10 \\ 8 & 5 \end{bmatrix} \begin{bmatrix} x \\ y \end{bmatrix} = \begin{bmatrix} 5.667 \\ 7.167 \end{bmatrix}$$

We solve and find $x = 0.27777$ or (5/18) $y = 0.72222$ or 13/18.

Example 2: Management-Labor Arbitration (Straffin, 2004 p 115-117)

			Labor Concedes			
		Nothing	C	A	CA
	Nothing	(0,0)	(4,-1)	(4,-2)	(8,-3)
Management Concedes	P	(-2,2)	(2,1)	(2,0)	(6,-1)
	R	(-3,3)	(1,2)	(1,1)	(5,0)
	PR	(-5,5)	(-1,4)	(-1,3)	(3,2)

Table 16. Management-versus Labour problem

The convex polygon is graphed from the constraints below (see the plots in figures 4):

$x + y \geq 0$
$0.5x + y \geq 0$
$0.25x + y \geq -1$
$x + y \geq 5$
$0.5x + y \leq 3.5$
$0.25x + y \leq \frac{15}{4}$

The status quo point (our security level) is (0,0), making the function to maximize simply $x^* y$.

Our formulation is:

Maximize x*y

Subject to::

$x + y \geq 0$
$0.5x + y \geq 0$
$0.25x + y \geq -1$
$x + y \geq 5$
$0.5x + y \leq 3.5$
$0.25x + y \leq \dfrac{15}{4}$

The product is $xy=6$ and the values are $x=3$ and $y=2$.

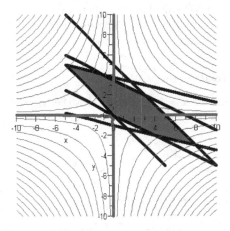

Figure 4. The graphical NLP problem for the Management-Labor Arbitration.

This optimal point is the point (3,2) on the line that is tangent to the contours in the direction of the NE increase.

10. N–person games

We restrict our discussion to the three person games. We suggest placing the payoffs into payoff matrices as shown in Table 17. We will continue to use Rose and Colin but introduce Larry as our third player. We show with only two strategies each but the concept can be expanded.

		Larry, L_1			Larry, L_2	
		Colin			Colin	
		C1	C2		C1	C2
Rose	R1	(r1,c1,l1)	(r1,c2,l1)	R1	(r1,c1,l2)	(r1,c2,l2)
	R2	(r2,c1,l1)	(r1,c2,l1)	R2	(r2,c1,l2)	(r2,c2,l2)

Table 17. Three-person game.

Again if $r_i + c_i + l_i = 0$ or the same constant for all i we have a total sum game otherwise we have a partial sum game.

Movement diagram may again be used to examine the game for pure strategy solution. Arrow point from the small values to the larger values. The new arrows belong to Larry. Between Larry 1 and Larry 2 we draw arrows from smaller to larger by an arrow out from one matrix and an arrow in to the other. We will illustrate with an example. Regardless if there is a pure solution or solutions or not, we will still consider coalitions. A coalition will be one two players joins together to gain an advantage of a third player. We consider all such coalition in our analysis.

Example Three Person Total Conflict

Consider the following three person total conflict game between Rose, Colin, and Larry. We provide the payoffs and the movement diagram with all arrows.

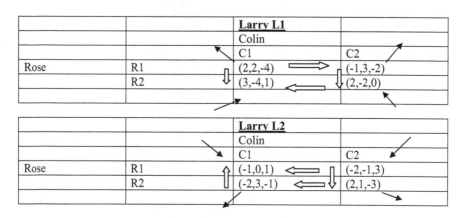

Table 18. Movement diagram for our three person game example.

Our movement arrows indicate two stable pure strategies, R2C1L1 (3,-4,1) and R1C1L2 (-1,0,1). These results are very different and not all players are satisfied at one or the other points. We now consider coalitions. We completely illustrate one coalition and provide the results for the others.

Let's assume that Larry and Colin form a coalition against Rose. Our new payoff matrix would look as follows:

Rose		Colin & Larry			
		C1L1	C2L1	C1L2	C2L2
	R1	2,-2	-1,1	-1,1	-2,2
	R2	3,-3	2,-2	-2,2	2,-2

Table 19. Colin and Larry coalition payoffs.

As a zero-sum game we may just list Rose's values.

Rose		Colin & Larry			
		C1L1	C2L1	C1L2	C2L2
	R1	2	-1	-1	-2
	R2	3	2	-2	2

Table 20. Colin and Larry coalition payoffs as a zero-sum game.

We can use linear programming to obtain our solution for Rose. Since payoffs are negative so that we solution can negative we employ the transformation of V_r to

$$V_{r1} - V_{r2}.$$

Maximize $V_{r1} - V_{r2}$

Subject to:

$$2y_1 - y_2 - y_3 - 2y_4 - V_{r1} - V_{r2r} \geq 0$$
$$3y_1 + 2y_2 - 2y_3 + 2y_4 - V_{r1} - V_{r2} \geq 0$$
$$y_1 + y_2 + y_3 + y_4 = 1$$
$$y_i, \ V_{r1} - V_{r2} \geq 0$$

We find the optimal solution is $V_{r1}=0$, $V=1.2$ so $V_r=-1.2$ when $y_1=0$, $y_2=0$, $y_3=4/5$ and $y_4=1/5$. Thus, the coalition of Colin-Larry gains 1.2 units where we find Larry get 21/25 of the share and Colin gets 9/25 of the share.

For the other coalitions, we may use the same procedures. Also, we may use the same procedures if we have a three person partial conflict game. However, for those coalitions we must use the (M,N) formulations since M+N does not equal zero.

Author details

William P. Fox
Naval Postgraduate School, Monterey, California, USA

References

[1]   Aumann, R. J. (1987). Game theory. The New Palgrave: A Dictionary of Economics, London, UK.: Palgrave Macmillan.

[2]   Bazarra, M. S., H. D. Sherali, and C. M . Shetty. (2006). Nonlinear Programming. 3rd Ed. New York, NY: John Wiley.

[3]   Braun, S. J. (1994). Theory of moves, American Scientist, 81, 562-570.

[4]   Camerer, C. (2003). Behavioral game theory: experiments in strategic interaction, Russell Sage Foundation, Description and Introduction, 1– 25.

[5]   Chiappori, A., Levitt, S.& Groseclose, P . (2002).Testing mixed-strategy equilibria when players are heterogeneous: the case of penalty kicks in soccer., American Economic Review, 92(4), 1138-1151.

[6]   Crawford, V. (1974). Learning the optimal strategy in a zero-sum game. Econometrica 42(5), 885–891.

[7]   Danzig, G. (1951). Maximization of a linear unction of variables Subject to: linear inequalities. In T.Koopman (Ed), Activity Analysis of Production and Allocation Conference Proceeding (pp. 339–347). New York, NY. John Wiley.

[8]   Danzig, G. (2002). Linear programming, Operations Research, 50(1), 42–47.

[9]   Dixit, A. & Nalebuff, B. (1991). Thinking strategically: The competitive edge in business, politics, and everyday life. New York, NY: W.W. Norton.

[10] Dorfman, Robert. (1951). Application of the simplex simplex method method to a game game theory theory problemproblem. In T. Koopman (Ed), Activity Analysis of Production and Allocation Conference Proceeding. (pp. 348–358). New York, NY. John Wiley Publishers. pp.339-347.

[11] Dutta, P. (1999). Strategies and games: theory and practice, Cambridge, MA.: MIT Press.

[12] Fox, W. P. (2008). Mathematical modeling of conflict and decision making: The writer's guild strike 2007–2008, Computers in Education Journal, 18(3), 2–11.

[13] Fox, W. P. (2012). Mathematical modeling with Maple. Boston, MA.: Cengage Publishers.

[14] Fox, W.P. (2010). Teaching the applications of optimization in game theory's zero-sum and non-zero sum games, International Journal of Data Analysis Techniques and Strategies, 2(3), 258-284.

[15] Gale, D., H. Kuhn, and A. Tucker. (1951). Linear programming and the theory of games. In T. Koopman (Ed), Activity analysis of production and allocation conference proceeding (pp. 317–329), New York, NY: John Wiley.

[16] Gillman, R., & Housman, D. (2009). Models of conflict and cooperation. Providence, RI.: American Mathematical Society.

[17] Gintis, H. (2000). Game theory evolving: a problem-centered introduction to modeling strategic behavior. Princeton, N.J.: University Press.

[18] Giordano, F., Fox, W., & Horton, S. (2013) A First Course in Mathematical Modeling, 5th Ed. Boston, MA.: Brooks-Cole.

[19] Harrington, J. (2008). Games, strategies, and decision making. New York, NY: Worth Publsihers.

[20] Isaacs, R. (1999). Differential games: A mathematical theory with applications to warfare and pursuit, control and optimization. New York, NY. Dover Publications.

[21] Klarrich, E. (2009). The mathematics of strategy. Classics of the Scientific Literature. October 2009 from www.pnas.org/site/misc/classics5.shtml.

[22] Koopman (Ed), Activity Analysis of Production and Allocation Conference Proceeding (pp. 348–358). New York, NY. John Wiley.

[23] Kuhn, H. W., and A.W. Tucker. (1951). Nonlinear programming. Proceedings of the Second Berkley Symposium on Mathematical Statistics and Probability, J. Newman (Ed.). Berkeley, CA.: University of California Press.

[24] Leyton-Brown, K.& Shoham, Y (2008), Essentials of game theory: A concise, multidisciplinary introduction, San Rafael, CA.: Morgan & Claypool Publishers.

[25] Miller, J. (2003). Game theory at work: how to use game theory to outthink and outmaneuver your competition, New York, NY.: McGraw-Hill.

[26] Myerson, Roger B. (1991). Game theory: analysis of conflict. Cambridge, MA.: Harvard University Press.

[27] Nash, J. (1950). The bargaining problem, Econometrica, 18, 155–162.

[28] Nash, J. (1951). Non-cooperative games, Annals of Mathematics, 54, 289–295.

[29] Nash, J. (2009). Lecture at NPS. Feb 19, 2009.

[30] Nash, John (1950), Equilibrium points in n-person games", Proceedings of the National Academy of Sciences of the United States of America. 36 (1): 48–49.

[31] Osborne, M. (2004). An introduction to game theory. Oxford, UK: Oxford University Press,.

[32] Papayoanou, P. (2010., Game theory for business, e-book, Probabilistic Publishing, http://www.decisions-books.com/Links.html.

[33] Rasmusen, Eric (2006). Games and information: An introduction to game theory. (4th ed.), New York, NY: Wiley-Blackwell .

[34] Shoham, Y. &; Leyton-Brown, K. (2009). Multiagent systems: Algorithmic, game-theoretic, and logical foundations. New York, NY: Cambridge University Press.

[35] Smith, J.M. & Price, G. (1973). The logic of animal conflict, Nature 246, 5427, 15–18.

[36] Smith, M. (1982). Evolution and the theory of games. Cambridge, UK: Cambridge University Press.

[37] Straffin, P. (1980). The prisoner's dilemma. UMAP Journal. 1, 101-113.

[38] Straffin, P. (1989). Game theory and nuclear deterrence, UMAP Journal. 10, 87-92.

[39] Straffin, P. D. (2004). Game theory and strategy. Washington, DC: Mathematical Association of America.

[40] Von Neumann, J., & Morgenstern, O. (2004). Theory of games and economic behavior (60th anniversary ed.). Princeton, NJ.: Princeton University Press.

[41] Webb, James N. (2007). Game theory: decisions, interaction and evolution. London, UK: Springer .

[42] Williams, J. D. (1986). The Compleat Strategyst. New York, NY: Dover Press.

[43] Winston, W. L. (2003). Introduction to Mathematical Programming. 4th Ed. Belmont, CA: Duxbury Press.

Chapter 3

Comparisons of Lateral Transshipment with Emergency Order Policies

Yi Liao, Wenjing Shen, Xinxin Hu and Benjamin Lev

Additional information is available at the end of the chapter

1. Introduction

The retail industry has been puzzled by stock-outs for a long time. According to a study report from Supply Chain Digest January 20, 2009, averagely, "more than 1 in every 5 consumers (21.2%) coming into the door of Consumer Electronics retailers leaves without buying at least one product they intended to purchase due to out-of-stocks". For example, Office Max has an out-of-stock rate of 30.6% and is losing $1.96 for every customer coming through their doors due to this reason.

If stock-out occurs, retailers often put emergency orders to meet customer's extra demand. For example, it is very common that oversee employees work over time to fulfill additional orders. On the other hand, transshipment is also a practical business solution to this problem. In the United States, it is commonly observed that if a customer goes to a car dealership and wants a certain type of car, and if the desired car (such as red color) is not in stock, the car dealership will arrange transshipment with another car dealer somewhere in the country with the exact car that the customer wants.

Though transshipment and emergency order problems have been addressed in many perspectives, it is quite rare that two policies are investigated at the same time in a comparative framework, especially with customer requesting behavior and customer switching behavior absorbed. In our research, customer requesting behavior describes that customers who don't acquire their desired products may submit requests to the retailer to ask for being satisfied by emergency orders or transshipments. Meanwhile, customer switching presence refers that some unmet customers may directly switch to another store to search the possibilities of shopping instead of requesting.

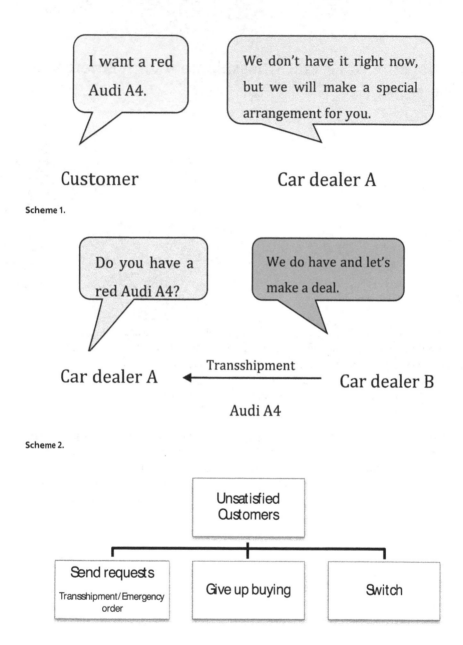

Figure 1. Unsatisfied Customers' Behaviors

In this paper, we study the transshipment and emergency order policies in the presence of "customer requesting" and "customer switching" behaviors, for two retailers under centralized control in a symmetric market system. As obtaining the correct stock balance gives the firm a competitive advantage, we first examine retailers' replenishment decision since in our model, customer demand randomly distributes before selling season. Considering customer requesting and switching factors, we are interested in how those initial inventory decisions should be adjusted correspondingly under two different polices(e.g., transshipment & emergency order). Through numerical experiments, we illustrate that retailer under transshipment policy usually needs to reserve more stock.

Secondly, we contrast the total supply chain's profits in our new model under two policies and aim to find convenient policy-choosing criteria. Under emergency order scenario, any switching customer satisfied by the surplus definitely improves the overall system's profit since the revenue is generated without any additional cost. In the meantime, firm using transshipment as the primary practice to solve out-of-stock issue can also benefit significantly from customer switching behavior by saving transshipment cost. Therefore, there is no straight forward conclusion for retailers regarding profit. We identify that with the same initial replenishment stock, in a symmetric scenario, retailer gains more if emergency cost is less than transshipment charge.

2. Related literature

Emergency order and transshipment, as effective solutions for increasing the multi-echelon supply chain performances have been given attention tremendously.

On one side, a number of models in the literature address models in which there is an option to place new emergency orders if shortage happens. The emergency order often has negligible lead-time, but the unit price is much more expensive. Daniel (1963) studies the optimality of periodic review order-up-to inventory policies when lead time is either 0 or 1 period. Later, Moinzadeh and Nahimas (1988) use a continuous review paradigm to develop a general heuristic policy. The emergency ordering procedure is triggered once on-hand inventory reaches a certain level. Under periodic review inventory system, Chiang and Gutierrez (1998) provide the optimal control policies at each review time point. Other studies can be found in Jain et al. (2010), Lawson and Porteus (2000), and Gaukler et al. (2009).

Although transshipment problem is analyzed in many different perspectives, we only review the research works which are closely related to our paper, where transshipments are conducted after customer demand is realized. Krishnan and Rao (1965) may be the first to explore a single-period two-location problem and its N location extension. Robinson (1990) considers a multi-period, multi-location problem where products are relocated among different locations. Under the assumption of zero transshipment and replenishment lead times, Robinson (1990) derives the optimal ordering policy and finds analytical solutions for the two-location case. Later, Herer and Rashit (1999) consider fixed joint replenishment costs in the similar model. Hu et al. (2008) study multiple period setting but focus on two-location transshipment. Most recently,

Olsson (2010) claims that a unidirectional lateral transshipment policy is reasonable if the locations have very different backorder or lost sales costs.

Most early studies assume that there exists a centralized inventory planer coordinating the optimal inventory and transshipment. However, Rudi et al. (2001) initially considers a two retailer decentralized one-period system and proves the uniqueness of Nash equilibrium in order quantities. Hu et al. (2007) discuss the existence of the coordinating transshipment prices. Huang and Sosic (2010) study a repeated inventory sharing game with N retailers and profit of transshipment is distributed among retailers by dual allocation. Other recent transshipment studies can be found in Yu et al. (2011) and Tiacci (2011).

As far as we know, studies mentioned so far all assume that unfilled demand of one retailer never turns to the other retailer. However, it is quite usual that consumers may simply leave and go for shopping at another retailer. In the existing literature, only few studies address the lateral transshipment problem and emergency ordering policy by explicitly incorporating such consumer switching behavior. Lippman and McCardle (1997) implement a rule to split initial and excess demand among competing firms in a competitive newsboy model. Anupindi and Bassok (1999) explore a one manufacturer and N-Retailer system, where a deterministic customer switching rate is assumed, and illustrate that the manufacturer may prefer a decentralized system when market search is intense. Other papers which explore customer switching behavior can be found in Jiang and Anupindi (2010) and Zhao and Atkins (2009). Although demand spills between firms are considered in those studies, transshipment issue and emergency order policy are never studied.

3. Model

In this study, we consider a centralized newsboy model consisting of two local retailers i, j facing their stochastic demands D_i, D_j independently. Before customer comes, the central controller decides the replenishment inventory levels Q_i, Q_j for retailer i, j. Then, the random demands D_i, D_j are realized and retailers use products on hand to accommodate their own customers respectively. As the local demand is very unpredictable, it is not surprising that retailers may not fulfill all the orders solely by their local inventories. Thereafter, in our model, we anticipate that a fraction of unsatisfied customers are willing to request retailer's transshipment/emergency order arrangement. Other unmet customers may immediately head for other stores and see if they can get their desired products or give up purchasing completely.

Supposing retailer i runs out of products and retailer j's inventory is adequate enough, we assume that $\lambda_i(D_i - Q_i)$ customers see whether transshipment or emergency order can be arranged for them, and remain at retailer i unless their requests are finally rejected, where we refer to the constant fraction parameter λ_i as "customer-requesting rate". Among the rest unmet demand $(1 - \lambda_i)(D_i - Q_i)$, the proportion of customers moving to retailer j instead of leaving directly is A_i. Though customer switching behavior may be influenced by a number of factors, such as distance between stores, availability of substitutable products, or access to inventory information, etc, we still can expect that consumer populations from same areas have relatively stable switching rate. Due to this reason, it is appropriate to consider a fixed portion of unsatisfied customers will be triggered to switch by out-of stock issue. At the end of the period, if switching customers still cannot get satisfied, they eventually leave without buying.

Under emergency order setting, we use q_i, q_j to represent the emergency orders placed by retailer i, j respectively. Stick to the same assumption $D_i - Q_i > 0$, $Q_j - D_j > 0$, retailer $j's$ surplus inventory is $Q_j - D_j > 0$, $q_j = |\min(Q_j - D_j - A_i(1 - \lambda_i)(D_i - Q_i), 0)|$ and $q_i = \lambda_i(D_i - Q_i)$. It is clear that retailer j places emergency orders only if retailer j cannot utilize its surplus to meet all switching customers. However, if the surplus at market retailer j is far beyond the number of switching demand, left stocks at retailer j may cause certain level of overall inefficiency since those products aren't used at all.

Different from emergency order policy, transshipment policy can be regarded as an internal way to enhance supply chain efficiency because no external resource is available in one period. When stock-out happens to both retailers, no transshipment will be conducted. While retailer j has surplus $Q_j - D_j > 0$, $q_{ji} = \min(Q_j - D_j - A_i(1 - \lambda_i)(D_i - Q_i), \lambda_i(D_i - Q_i))^+$ is relocated from retailer j to retailer i, in responding to those customers' requests. Clearly, it is not necessary all customers who stay at local retailer i have to be satisfied by transshipment since partial extra products at retailer j are prepared for switching customers because of saving transportation cost. The extreme case occurs when the quantity of switching customers is large enough to meet all left products at retailer j, where retailers end up with no transshipment.

At this moment, we have briefly introduced our research model where retailers can choose one of two alternatives to handle demand uncertainty. With the help of transshipment, firm definitely can take advantage of customer switching behavior by saving shipping cost. Unfortunately, transshipment never meets all customers once out-of-stock takes place since no new merchandises are brought in.

Emergency order policy is also not a perfect substitute because possible waste may be incurred as mentioned early. In the following, our research first considers two policies separately, addressing on some critical operations management decisions, for example replenishment decision in a more realistic model. Furthermore, we pay attention to comparison of two policies and suggest how to decide the optimal policy under different parameter assumptions.

In our research, the regular unit inventory cost is c_n and retailers receive revenue $r > c_n$ for each unit sold locally as well as to switching customers. For each unit of inventory transshipped from retailer i to retailer j, a transshipment expense $t_{ij} < r$ is incurred. Manufacturer charges any emergency order $c_e > c_n$. We summarize the parameters used in our general model below.

Summary of notations

c_n = unit regular product cost;

c_e = unit emergency order cost;

$r =$ unit retail price;

$t_{ij} =$ unit transshipment cost from retailer i to retailer j;

$D_i =$ local demand at retailer i, $cdf = G_i(D_i)$ and $pdf = g_i(D_i)$;

$\lambda_i =$ requesting rate of retailer i's customer;

$A_i =$ switching rate of retailer i's customer;

4. Inventory policies under emergency order Policy

In order to optimize the total profit, the central controller has to plan on replenishment inventory level by minimizing the cost of stocks while trying to make sure that there are enough materials to meet customer demand. Firstly, this research displays an emergency order quantity schedule, and then investigates the consequences of customer switching and requesting behaviors. Finally, we extensively discuss the properties of the optimal replenishment decisions.

4.1. Emergency order schedule

When local demands are perceived and satisfied by retailers' products on hand, the central planning firm needs to arrange emergency orders when it's necessary. Without loss of generality, we present the emergency order schedule in Table 1 and Figure 1.

Event	Description	q_j	q_i
$Event_1$	$Q_i \geq D_i,\ Q_j \geq D_j$	0	0
$Event_2$	$D_i \geq Q_i,\ Q_j \geq D_j$ $Q_j - D_j \geq A_i(1 - \lambda_i)(D_i - Q_i)$	0	$\lambda_i(D_i - Q_i)$
$Event_3$	$D_i \geq Q_i,\ Q_j \geq D_j$ $Q_j - D_j \leq A_i(1 - \lambda_i)(D_i - Q_i)$	$A_i(1 - \lambda_i)(D_i - Q_i)$ $-(Q_j - D_j)$	$\lambda_i(D_i - Q_i)$
$Event_4$	$Q_i \geq D_i,\ D_j \geq Q_j$ $Q_i - D_i \geq A_j(1 - \lambda_j)(D_j - Q_j)$	$\lambda_j(D_j - Q_j)$	0
$Event_5$	$Q_i \geq D_i,\ D_j \geq Q_j$ $Q_i - D_i \leq A_j(1 - \lambda_j)(D_j - Q_j)$	$\lambda_j(D_j - Q_j)$	$A_j(1 - \lambda_j)(D_j - Q_j)$ $-(Q_i - D_i)$
$Event_6$	$Q_i \leq D_i,\ Q_j \leq D_j$	$A_j(1 - \lambda_j)(D_j - Q_j)$ $+\lambda_j(D_j - Q_j)$	$A_i(1 - \lambda_i)(D_i - Q_i)$ $+\lambda_i(D_i - Q_i)$

Table 1. Emergency Order Schedule

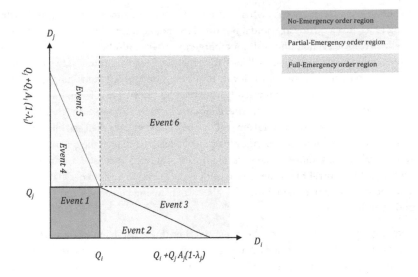

Figure 2. Emergency Order Structure

In Table 1, emergency order is not needed when $Event_1$ take places, since both retailers have surpluses. In Figure 1, this is labeled as a "No-Emergency order region". In $Event_4$ and $Event_5$, $\lambda_j(D_j - Q_j)$ unsatisfied customers wait for emergency orders provided by local retailer's arrangement and $(1 - \lambda_j)(D_j - Q_j)A_j$ customers will go for shopping. It is not hard to find that $Event_2$ and $Event_3$ are the counterparts of $Event_4$ and $Event_5$, e.g., $D_j < Q_j$, $D_i > Q_i$, which can be analyzed similarly by exchanging subscripts i with j.

In events $Event_2$ and $Event_4$, it is suggesting that retailers with extra products don't place any emergency order since extra stocks cover all switching customers. On the other hand, in $Event_3$ and $Event_5$, leftovers are not sufficient enough to serve all switching orders, which require retailers send emergency requests. Although events mentioned above except $Event_1$ are not exactly same, in general, all unmet customers are compensated by products mixed of emergency orders and surplus products. Because of this, we refer these regions as "Partial-Emergency order region" in Figure 1. Nonetheless, since in $Event_6$, emergency orders become the only available resource to solve out-of-stock problem, we label this region as a "Full-emergency order region". Since we are particularly interested in the effects of customer switching and requesting behaviors, we explain their impacts in Proposition 1.

Proposition 1 The amount of emergency orders $q_i + q_j$ is non-decreasing in customer requesting rate λ_i, λ_j, and customer switching rate A_i, A_j.

Recall that in emergency order problems without customer switching and requesting, emergency order decision and retailer's inventory surplus level are isolated from each other. The total amount of emergency orders is simply the sum of all unsatisfied demands from every single shortage retailer. However, when our model adopts customer requesting and customer switching rates, this rule doesn't hold anymore. First, not all unsatisfied customers are willing to stay with the local retailer and wait for emergency orders. Additionally, for non-requesting customers, only a fraction of them will look for products. In the meantime, if two retailers have the opposite positions on inventory level, depending on the left stock amount, switching customers may be partially or completely absorbed by the surplus. Therefore, we conclude that in our model, the total amount of emergency orders never surpasses the number of unmet customers. After we explore the total emergency order amount, we then establish the findings of customer switching and requesting rates in the following. In fact, the sensitivities on rates are quite intuitive. As more customers choose to stay and request for emergency order arrangement, it is natural that more emergency orders need to be added. On the other hand, any increment in customer switching rate also avoids losing customers and more needs are satisfied overall.

Next, we emphasize on addressing the question: How does customers' preference on requesting or switching affect the profit measurement? Intuitively, both rates measure the extent of demand pooling between retailers. But, the profit performance depends on many factors, such as the retail price. Hence, we have,

Proposition 2 Under emergency order setting, the total profit increases in customer switching rate. When $rA_j - r + c_e < 0$, then the total expected profit increases in λ_j.

Compared with requesting rate, customer switching behavior's impact on system's profit is relatively obvious. As analyzed before, the higher proportion of customer switching rate, the fewer customers leave with disappointment since switching customers eventually get fully satisfied by the surplus or emergency order, which helps the supply chain to achieve a better financial performance. At this moment, the retail price does not play a decisive role since switching customers only come from those who prefer not to wait for emergency order.

However, we cannot simply extend customer switching rate sensitivity conclusion to customer requesting rate. We still follow our original assumption that retailer i has surplus and retailer j is short of products. As customer requesting rate rises, more income is generated because of more waiting customers at retailer j. But, at the same time, fewer customers are expected to switch, which results in less revenue from retailer i. Particularly, when retailer i can use its surplus to meet all switching customers, this loss is more significant. Therefore, it is not straightforward that the gain from more waiting customers at retailer j make can make up the loss from fewer switching customers at retailer i without considering the retail price and emergency order cost. Analytically, if one unit of extra inventory at retailer i is sold at price r, the opportunity cost can be calculated as $rA_j - r + c_e$ because one unit of inventory may also be sold to any waiting customer at price r, emergency cost at retailer j is c_e, and expected revenue from retailer i is rA_j. Hence, only the benefit of waiting customers dominates the benefit of switching customers, it is worthwhile to encourage unsatisfied customers to stay.

4.2. Characteristics of optimal replenishment decisions

The replenishment decision needs to be made before demand realization. For any inventory replenishment levels Q_i, Q_j, the total expected profit is given by:

$$E\pi^E = -c_n(Q_i + Q_j) + \int_0^{Q_i}\int_0^{Q_j}\pi_1^E g_i(D_i)g_j(D_j)dD_idD_j + \int_{Q_i}^{\infty}\int_{Q_j}^{\infty}\pi_6^E g_i(D_i)g_j(D_j)dD_idD_j \tag{1}$$

$$+\int_0^{Q_i}\int_{Q_j}^{\frac{Q_i-D_i}{(1-\lambda_j)A_j}+Q_j}\pi_2^E g_i(D_i)g_j(D_j)dD_idD_j + \int_0^{Q_i}\int_{\frac{Q_i-D_i}{(1-\lambda_j)A_j}+Q_j}^{\infty}\pi_3^E g_i(D_i)g_j(D_j)dD_idD_j \tag{2}$$

$$+\int_0^{Q_j}\int_0^{\frac{Q_j-D_j}{(1-\lambda_i)A_i}+Q_i}\pi_4^E g_i(D_i)g_j(D_j)dD_idD_j + \int_0^{Q_j}\int_{\frac{Q_j-D_j}{(1-\lambda_i)A_i}+Q_i}^{\infty}\pi_5^E g_i(D_i)g_j(D_j)dD_idD_j \tag{3}$$

Note: π_i^E, $i=1, \dots 6$ is the net income under different conditions.

π_1^E	$rD_i + rD_j$
π_2^E	$r(D_i + (1-\lambda_j)A_j(D_j - Q_j)) + r(Q_j + \lambda_j(D_j - Q_j)) - c_e\lambda_j(D_j - Q_j)$
π_3^E	$r(D_i + (1-\lambda_j)A_j(D_j - Q_j)) + r(Q_j + \lambda_j(D_j - Q_j))$ $-c_e((1-\lambda_j)A_j(D_j - Q_j) - (Q_i - D_i) + \lambda_j(D_j - Q_j))$
π_4^E	$r(D_j + (1-\lambda_i)A_i(D_i - Q_i)) + r(Q_i + \lambda_i(D_i - Q_i)) - c_e\lambda_i(D_i - Q_i)$
π_5^E	$r(D_j + (1-\lambda_i)A_i(D_i - Q_i)) + r(Q_i + \lambda_i(D_i - Q_i))$ $-c_e((1-\lambda_i)A_i(D_i - Q_i) - (Q_j - D_j) + \lambda_i(D_i - Q_i))$
π_6^E	$rQ_i + rQ_j + (r - c_e)(\lambda_i(D_i - Q_i) + (1-\lambda_j)A_j(D_j - Q_j))$ $+(r - c_e)(\lambda_j(D_j - Q_j) + (1-\lambda_i)A_i(D_i - Q_i))$

Table 2. The Retailers' Net Income Structure under Emergency Order Policy

The first two items in (1) are inventory replenishment costs of regular orders; the third one is the expected revenue generated from local demands when both retailers have extra inventories, and the forth one is the expected income if out of stock problem happens. (2) is the expected income when retailer i has surplus and retailer j runs out of stocks. (3) is the counterpart of (2), where firms' inventory positions are opposite.

Proposition 3 For a given Q_j, denote the optimal Q_i that maximizes the total profit by $Q_i^*(Q_j)$ is unique and decreases in Q_j.

It is not shocking that Q_i^* is declining because extra inventory at retailer j reduces the firm's stock-out risk when retailer i underestimates the level of demand for its products and increases its overstocking risk when retailer i has surplus, thus less inventory is needed. This is different from traditional emergency order problems, where one unit of increment at retailer j has no effect on retailer i's replenishment level. The difference is mainly driven by customer switching behavior.

Furthermore, we extend our research by identifying the intersection of $Q_i^*(Q_j)$ and $Q_j^*(Q_i)$ and explore the existence of the optimal pair (Q_i^E, Q_j^E). Because we introduce customer switching factor and customer requesting issue in our model, its concavity in (Q_i, Q_j) is not as obvious as before.

Theorem 1 The expected total profit $E\pi^E$ is jointly concave in (Q_i, Q_j). Therefore, there exists a unique pair of replenishment levels (Q_i^E, Q_j^E) that maximizes the total expected profit.

Although it is not easy to predict the retailer's replenishment decision's general trend, we can establish some sensitivity conclusions based on the following results.

Proposition 4 For a given Q_j, denote the optimal Q_i that maximizes the total profit by $Q_i^*(Q_j)$ increases in A_j and decreases in λ_j.

The sensitivity on replenishment level is quite intuitive. Besides this, Proposition 4 also describes the relationship between two retailers' replenishment levels when the system deviates from symmetric parameter setting, which provides some managerial insights for retailers.

5. Inventory policies under transshipment policy

We now turn to study the case where a central coordination scheme controls the inventory and transshipment is implemented when it is necessary.

5.1. Transshipment schedule

Similarly, we first check the transshipment decisions after demands at both retailers are observed. Note that transshipment happens only if two retailers have different stock inventory statuses, i.e., one runs out of products and the other one hold too much stock. Below, adopting the framework from Table1, we exhibit the complete transshipment schedule in Table 3

From Table 2, in $Event_1$ and $Event_8$, there is no transshipment since retailer i and retailer j are overstocking their products. In $Event_8$, customer switching behavior doesn't matter since no unsatisfied customer will be served. This is totally different from the same case when retailer utilizes emergency order policy, where all switching demands are finally fulfilled by some

Event	Description	q_{ij}	q_{ji}
$Event_1$	$Q_i \geq D_i, \ Q_j \geq D_j$	0	0
$Event_2$	$Q_i \geq D_i, \ D_j \geq Q_j$ $Q_i - D_i \geq (A_j(1-\lambda_j)$ $+\lambda_j)(D_j - Q_j)$	$\lambda_j(D_j - Q_j)$	0
$Event_3$	$Q_i \geq D_i, \ D_j \geq Q_j$ $Q_i - D_i \geq A_j(1-\lambda_j)(D_j - Q_j)$ $Q_i - D_i \leq (A_j(1-\lambda_j)$ $+\lambda_j)(D_j - Q_j)$	$Q_i - D_i - A_j(1-\lambda_j)$ $(D_j - Q_j)$	0
$Event_4$	$Q_i \geq D_i, \ D_j \geq Q_j$ $Q_i - D_i \leq A_j(1-\lambda_j)(D_j - Q_j)$	0	0
$Event_5$	$D_i \geq Q_i, \ Q_j \geq D_j$ $Q_j - D_j \geq (A_i(1-\lambda_i) + \lambda_i)$ $(D_i - Q_i)$	$\lambda_i(D_i - Q_i)$	0
$Event_6$	$D_i \geq Q_i, \ Q_j \geq D_j$ $Q_j - D_j \geq A_i(1-\lambda_i)(D_i - Q_i)$ $Q_j - D_j \leq (A_i(1-\lambda_i) + \lambda_i)$ $(D_i - Q_i)$	0	$Q_j - D_j - A_i(1-\lambda_i)$ $(D_i - Q_i)$
$Event_7$	$D_i \geq Q_i, \ Q_j \geq D_j$ $Q_j - D_j \leq A_i(1-\lambda_i)(D_i - Q_i)$	0	0
$Event_8$	$Q_i \leq D_i, \ Q_j \leq D_j$	0	0

Table 3. Transshipment Schedule

external emergency orders. Due to this reason, we call emergency order policy "External Policy" versus transshipment policy "Internal Policy".

If we take a look at the transshipment table carefully, there are two other events i.e., $Event_4$ and $Event_7$ ending up with no transshipment since the extras at the surplus retailers are less than the switching demands and there is no reason to arrange any product relocation. Such transshipment pattern is primarily induced by customer switching behavior and is completely unlike classic research. It can also be deduced from Table 2 that in $Event_5$ and $Event_6$, the optimal transshipment quantity is $q_{ji} = \min\left(Q_j - D_j - A_i(1-\lambda_i)(D_i - Q_i), \ \lambda_i(D_i - Q_i)\right)^+$, $x^+ = \max(x, 0)$, where $D_j < Q_j$ and $D_i > Q_i$, which are counterparts of $Event_2$ and $Event_3$. The logic behind the schedule is that if the surplus is more than the sum of switching and requesting customers, it is always optimal to transship as much as necessary. In contrast, when the surplus inventory is not enough to match all potential shortfalls, it is better to meet switching customers first, and then transport leftovers.

In the same vein, we also examine how customer switching and requesting behaviors impact the transshipment quantity. The following proposition summarizes some straightforward effects.

Proposition 5 The transshipment amount q_{ij} is increasing in shipping-request rate λ_j and non-increasing in customer switching rate A_j.

Although transshipment policy is quite different from emergency order policy, their structures appear great similarity. Then we have,

Proposition 6 The total profit is increasing in customer switching rate. When $rA_j - r + t_{ij} > 0$, then total expected profit is decreasing in λ_j.

The major difference between Proposition 2 and Proposition 6 is that c_e in Proposition 2 is replaced by t_{ij}. However, we still can develop the similar marginal analysis for this deviation. For each unit inventory relocated from retailer i to retailer j, it is priced at transportation charge t_{ij} and yields revenue r. Meanwhile, averagely we can regard rA_j as the return brought in by any single switching customer. When $rA_j - r + t_{ij} < 0$, the whole system loses money if customers prefer waiting instead of switching.

5.2. Characteristics of optimal replenishment decisions under transshipment policy

We then move to study the replenishment decisions under transshipment setting. Under central coordination, we choose Q_i, Q_j to maximize the total expected profit which is given by:

$$E\pi^T = - c_n(Q_i + Q_j) + \int_0^{Q_i}\int_0^{Q_j}\pi_1^T g_i(D_i)g_j(D_j)dD_idD_j + \int_{Q_i}^{\infty}\int_{Q_j}^{\infty}\pi_8^T g_i(D_i)g_j(D_j)dD_idD_j \tag{4}$$

$$+\int_0^{Q_i}\int_{Q_j}^{\frac{Q_i-D_i}{(1-\lambda_j)A_j+\lambda_j}+Q_j}\pi_2^T g_i(D_i)g_j(D_j)dD_idD_j + \int_0^{Q_i}\int_{\frac{Q_i-D_i}{(1-\lambda_j)A_j}+Q_j}^{\frac{Q_i-D_i}{(1-\lambda_j)A_j+\lambda_j}+Q_j}\pi_3^T g_i(D_i)g_j(D_j)dD_idD_j \tag{5}$$

$$+\int_0^{Q_i}\int_{\frac{Q_i-D_i}{(1-\lambda_j)A_j}+Q_j}^{\infty}\pi_4^T g_i(D_i)g_j(D_j)dD_idD_j + \int_0^{Q_i}\int_0^{\frac{Q_j-D_j}{(1-\lambda_i)A_i+\lambda_i}+Q_i}\pi_5^T g_i(D_i)g_j(D_j)dD_idD_j \tag{6}$$

$$+\int_0^{Q_i}\int_{\frac{Q_j-D_j}{(1-\lambda_i)A_i+\lambda_i}+Q_i}^{\frac{Q_j-D_j}{(1-\lambda_i)A_i}+Q_i}\pi_6^T g_i(D_i)g_j(D_j)dD_idD_j + \int_0^{Q_i}\int_{\frac{Q_j-D_j}{(1-\lambda_i)A_i}+Q_i}^{\infty}\pi_7^T g_i(D_i)g_j(D_j)dD_idD_j \tag{7}$$

Note: π_i^T, $i = 1, \ldots 8$ is the net income under different conditions.

π_1^T	$rD_i + rD_j$
π_2^T	$r(D_i + (1 - \lambda_j)A_j(D_j - Q_j)) + r(Q_j + \lambda_j(D_j - Q_j)) - t_{ij}\lambda_j(D_j - Q_j)$
π_3^T	$r(D_i + (1 - \lambda_j)A_j(D_j - Q_j)) + r(Q_j + (Q_i - D_i) - (1 - \lambda_j)A_j(D_j - Q_j))$ $-t_{ij}((Q_i - D_i) - (1 - \lambda_j)A_j(D_j - Q_j))$
π_4^T	$rQ_i + rQ_j$
π_5^E	$r(D_j + (1 - \lambda_i)A_i(D_i - Q_i)) + r(Q_i + \lambda_i(D_i - Q_i)) - c_{ji}\lambda_i(D_i - Q_i)$
π_6^T	$r(D_j + (1 - \lambda_i)A_i(D_i - Q_i)) + r(Q_i + (Q_j - D_j) - (1 - \lambda_i)A_i(D_i - Q_i))$ $-t_{ij}((Q_j - D_j) - (1 - \lambda_i)A_i(D_i - Q_i))$
π_7^T	$rQ_i + rQ_j$
π_8^T	$rQ_i + rQ_j$

Table 4. The Retailers' Net Income Structure under Transshipment Policy

Proposition 7 For a given Q_j, denote the optimal Q_i that maximizes the total profit by $Q_i^*(Q_j)$ decreases in λ_j and increases in A_j if the demand density function g_i is an increasing function.

Though transshipment and emergency order policy are not identical, the optimal replenishment level possesses some common properties. The essential point is that these decisions are Newsvendor-type inventory model decisions. In the aspect of optimal inventory levels, we can show

Theorem 2 The expected total profit $\pi^T = (Q_i, Q_j)$ is jointly concave in (Q_i, Q_j). Therefore, there exists a unique pair of replenishment levels (Q_i^T, Q_j^T) that maximizes the expected total profit.

Theorem 1 and Theorem 2 exhibit that if two retailers have a common supplier, the optimal replenishment levels definitely can be achieved despite retailers prefer transshipment or emergency order policy. Since good inventory management will lower costs, improve efficiency, Theorem 1 and 2 also provide some operational insights for management.

We have analytically studied customer requesting and customer switching phenomena's impacts on the firm's operations under emergency order policy or transshipment respectively. Although choosing emergency order policy or transshipment is determined by firm's own preference, it is still worthwhile if we can provide some straightforward comparisons between two polices. Then we have,

Proposition 8 When two retailers are symmetric, if emergency order cost is less or equal to transshipment cost, i.e., $c_e < t_{ij}$, t_{ji}, emergency order policy always dominates transshipment policy in profitability with the same initial replenishment inventories.

The conclusion above gives us a simple, practical guideline when two markets are in a similar position. In order to make business more profitable, retailer can easily make the decision if transshipment cost requires more.

6. Computational analysis

In the analytical study part, we have proved that customer switching and requesting factors influence retailers' decisions substantially. In this part, we numerically evaluate these effects and especially interested in comparing the operations index such as replenishment level, profitability of two polices. Because our study focuses on a symmetric scenario, we drop the subscripts "i, j", which are used to distinguish retailers. In our experiments, customer demand is normally distributed with mean 100 and standard deviation 50. Before proceeding to our model's results, we first demonstrate the traditional model's numerical outcomes.

Then, by adjusting the values of λ, A respectively, we investigate the trends of the optimal replenishment levels under different polices in our new framework. Next, we fix the initial inventory amount and examine how profit varies corresponding to changes of requesting rate and customer switching rate. Before moving to the details, we list all notations used in this section below:

A = Customer switching rate

c_e = Emergency order cost

λ = Customer requesting rate

t = Transshipment cost

Q_T = Optimal replenishment level under transshipment

Q_E = Optimal replenishment level under emergency order

PT = Total profit under transshipment

PE = Total profit under emergency order

IPC = Inventory level changes from initial level

6.1. Observation 1

In the classical centralized model, unsatisfied customers at each retailer only wait for transshipment or emergency order and never switch or give up purchasing. We let the unit cost c_n = 20, retail price r = 40 and list the result in the following table.

Obviously, retailers using emergency order policy require less replenishment inventories and generate more profits. Since holding more stocks increase costs for businesses such as increases warehouse space needed, money spent in stocks could have been allocated etc, it sounds like

c_e	Q_E	PE	Q_T	PT	t
22	39.1	3951.9	113.5	2957.6	2
24	55.3	3837.3	113	2940.9	4
26	66	3744.6	112.5	2923.9	6
28	74.1	3666.2	112	2906.7	8
30	80.5	3598.1	111.4	2889.2	10
32	85.9	3537.9	110.8	2871.4	12
34	90.6	3484	110.2	2853.3	14
36	94.6	3435.2	109.7	2835	16
38	98.2	3390.6	109.1	2816.3	18
40	101.4	3349.6	108.5	2797.3	20

Table 5. Traditional Models' Numerical Results

that retailers should exert emergency order policy as often as possible. However, since the above model's assumption is far from the realistic business environment, we are more concerned about whether the supply chain still behaves in the similar way. In the following, we display our experimental outcome in our modified model.

6.2. Observation 2

We still fix the normal unit cost $c_n = 20$, retail price $r = 40$ and let $A = 0.2$. The emergency cost c_e is priced at 30 and transshipment price t is marketed at 30 and 2 respectively, which can exhibit two extreme cases: (1) no disparity in cost (2) large gap between two charges.

λ	Q_E	PE	$Q_{T\ t=30}$	$PT_{t=30}$	$IPC_{t=30}$	$Q_{T\ t=2}$	$PT_{t=2}$	$IPC_{t=2}$
0.1	96.2	2780.5	105.5	2722.9		120.6	2621.4	
0.2	94.8	2780	104.9	2726.5	0.64%	118.2	2692.1	1.99%
0.3	93.4	2777.5	104.2	2728.4	1.22%	115.7	2758	4.06%
0.4	91.9	2772.6	103.7	2728.6	1.75%	113.3	2817.5	6.05%
0.5	90.4	2765	103.2	2727.3	2.23%	110.9	2869.5	8.04%
0.6	88.7	2753.9	102.7	2724.5	2.66%	108.7	2913.7	9.87%
0.7	86.9	2738.9	102.3	2720.5	3.04%	106.6	2950	11.61%
0.8	84.9	2719.9	102	2715.3	3.36%	104.7	2979	13.18%

Table 6. $A = 0.2$

A	Q_E	PE	$Q_{T\ t=30}$	$PT_{t=30}$	$IPC_{t=30}$	$Q_{T\ t=2}$	$PT_{t=2}$	$IPC_{t=2}$
0.1	94.8	2704	105.7	2672.4		119.7	2631.8	
0.2	93.5	2780	104.9	2726.5	0.76%	118.2	2692.1	1.25%
0.3	92.3	2853.3	104.3	2775.5	1.32%	116.8	2746.9	2.42%
0.4	91.2	2923.6	103.9	2819.6	1.70%	115.6	2796.2	3.43%
0.5	90.2	2991	103.5	2859.1	2.08%	114.6	2840	4.26%
0.6	89.2	3055.6	103.5	2894.4	2.08%	113.8	2878.8	4.93%
0.7	88.3	3177.6	103.5	2954.2	2.08%	113.1	2913.1	5.51%
0.8	88.3	3177.2	103.6	2954.2	1.99%	112.6	2943.4	5.93%

Table 7. $\lambda = 0.2$

From Table 6 and Table 7, the inventory level under transshipment always dominates the one under emergency policy despite deviation between transshipment expense and emergency order cost is zero or huge. This discovery displays the same pattern as the traditional model. Intuitively, this justifies our statement mentioned before, where we call transshipment an internal method and emergency order policy an external method. Since retailer cannot get help from outside resource, it is not surprising that retailers conservatively hold more stock to alleviate out-of-stock problem.

But it is also notable that emergency order policy doesn't prevail transshipment policy anymore in profitability under our customer behavior absorbed inventory model framework, for example, $\lambda = 0.5$, $t = 2$, $A = 0.2$, $c_e = 30$, $PT = 2869.5$, $PE = 2727.3$. It advises retailers that they still can be more profitable without placing emergency orders, especially when transshipment cost is relatively low.

Though Table 6 exhibits the impact of customer requesting behavior on replenishment level is trivial when transshipment cost is expensive, this effect becomes more prominent as transshipment price declines. When more customers send requests, we anticipate that more unmet customers are willing to stay. As transshipment price is low, the marginal benefit is considerable if one transshipped unit is sold regarding the corresponding loss of over-stocking, which encourages firm to retain inventory more aggressively. On the other hand, as trans-shipment expense reaches a certain level, the marginal benefit is not that significant and firm reduces its pace of increasing stock reserving level.

In Table 7, although we conduct the sensitivity analysis on customer switching rate, the replenishment level change under transshipment displays a similar pattern. As customer switching scale rises, more customers are going for shopping instead of waiting for transship-ment. At this moment, transshipment price doesn't play a critical role. Due to this reason, it is reasonable that replenishment inventory level change rate doesn't vary rapidly.

6.3. Observation 3

In the previous numerical experiments, we focus on the optimal replenishment inventory levels which heavily depend on parameter settings. Nonetheless, in practice, retailers may simply order a certain number of products, since it is easily handled by employees. Because of this, in the experiments below, we lock the order and contrast the returns between two policies.

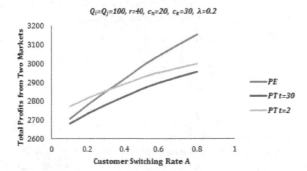

Figure 3. Retailers' Profits V.S Customer Switching Rate

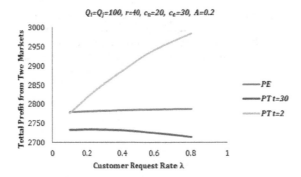

Figure 4. Retailers' Profits V.S Customer Requesting Rate

First of all, from Figure 3 and Figure 4, it is easily indentified that when transshipment cost decreases, the system's profit does increase dramatically. Therefore, firm may take advantage of this property and put effort on reducing transshipment expense since it does not need to consider customer's consumption behavior.

Secondly, we exhibit that as transshipment price is exactly same as emergency cost, emergency order policy completely outperforms transshipment method, which confirms the statement from Proposition 8. This result may be used as one of criteria for firm to choose the appropriate policy if retailers choose to order in a simple manner.

Lastly, when it costs firm high price to conduct transshipment, customer switching behavior and customer requesting rate may affect the total profit in the opposite direction. For example, in Figure 3, when transshipment cost is 30, it demonstrates an increasing trend as customer switching rate increases. In Figure 4, this trend reverses when customer requesting rate rises. The implication behind this is that high transshipment price may take a bite of the total profit if more customers are willing to stay and want for relocated products. Under this condition, retailer may facilitate customers to switch instead of waiting. On the other hand, when transshipment cost is relatively low, customers are more welcome to stay since it enhances the total profit considerably.

7. Summary and future research

This paper utilizes a centralized model to investigate how customers' switching and requesting behaviors affect retailers' operations decisions under emergency order and transshipment policies respectively. We first prove that the optimal replenishment stocking levels exist under two policies. Then, we explore how customer switching and shipping impact transshipment amount and emergency quantity. Furthermore, we numerically compare the system's performance in profitability under two policies, with /without the optimal replenishment level and provide some practical policy choosing criteria for retailers.

Since in our research we consider a centralized one-period two-location model, there are a lot of possible extensions for the future research. At the first step, we may expand our research to a decentralized setting, where retailer maximizes its own profit instead of overall performance. It's also worthwhile to extend our one-period two-location to N-period and N-location model among retailers.

Author details

Yi Liao[1*], Wenjing Shen[2], Xinxin Hu[3] and Benjamin Lev[2]

*Address all correspondence to: yiliao@swufe.edu.cn

1 Southwestern University of Finance and Economics, China

2 Drexel University, USA

3 Indiana University, USA

References

[1] Anupindi, R, & Bassok, Y. Centralization of stocks: retailers vs. manufacturer. Management Science. (1999). , 45(1999), 178-191.

[2] Chiang, C, & Gutierrez, G. J. Optimal control policies for a periodic review inventory system with emergency orders. Naval Research Logistics Quarterly. (1998). , 45(1998), 187-204.

[3] Daniel, K. H. A delivery-lag inventory model with emergency order. Stanford University Press, Stanford, CA, (1963).

[4] Gaukler, G. M, Ozer, O, & Hausman, W. H. Order progress information: Improved dynamic emergency ordering policies. Production and Operations Management. 17(6) ((2009).

[5] Herer, Y, & Rashit, A. Lateral Stock Transshipment in a two-location inventory system with fixed or joint replenishment costs. Naval Research Logistics. (1999). , 46(1999), 525-548.

[6] Hu, X, Duenyas, I, & Kapuscinski, R. Existence of coordinating transshipment prices in a two-location inventory model. Management Science. (2007). , 53(2007), 1289-1302.

[7] Hu, X, Duenyas, I, & Kapuscinski, R. Optimal Joint Inventory and Transshipment under Uncertain Capacity. Operations Research, (2008). , 56(2008), 881-897.

[8] Huang, X, & Sosic, G. Repeated news-vendor game with transshipment under dual allocations, European Journal of Operational Research. (2010). , 204(2010), 274-284.

[9] Jain, A, Groenevelt, H, & Rudi, N. Continuous review inventory model with dynamic choice of two freight modes with fixed costs. Manufacturing and Service Operations Management. (2010). , 12(2010), 120-139.

[10] Jiang, L, & Anupindi, R. Customer-Driven vs. Retailer-Driven Search: Channel Performance and Implications. Manufacturing and Service Operations Management. (2010). , 12(2010), 102-119.

[11] Krishman, K. S, & Rao, V. R. K. Inventory control in n warehouses. Journal of Industrial Engineering. (1965). , 16(1965), 212-215.

[12] Lawson, D, & Porteus, E. Multistage inventory management with expediting. Operations Research. 48(6) ((2000).

[13] Lippman, S, & Mccardle, K. The competitive newsboy. Operations Research. (1997). , 45(1997), 54-56.

[14] Moinzadeh, K, & Nahmias, S. A continuous review inventory model for an inventory system with two supply modes. Management Science. (1988). , 34(1988), 761-773.

[15] Olsson, F. An inventory model with unidirection lateral transshipment. European Journal of Operational Research. (2010). , 200(2010), 725-732.

[16] Robinson, L. W. Optimal and approximate policies in multi-period multi period inventory models with transshipments. Operations Research, (1990). , 38(1990), 278-295.

[17] Rudi, N, Kapur, S, & Pyke, D. F. A two-location inventory model with transshipment and local decision making. Management Science. (2001). , 47(2001), 1668-1680.

[18] Tiacci, L, & Saetta, S. A heuristic for balancing the inventory level of different locations through lateral transshipment. International Journal of Production Economics. (2011). , 131(2011), 87-95.

[19] Yu, D. Z, & Tang, S. Y. Niederhoff. On the benefits of operational flexibility in a distribution network with transshipment. Omega.(2011). , 39(2011), 350-361.

[20] Zhao, X, & Atkins, D. Transshipment between Competing Retailers. IIE Transactions. (2009). , 41(2009), 665-676.

Chapter 4

Managing Pharmacy Operations with People and Technology

Margaret O. Afolabi and
Omoniyi Joseph Ola-Olorun

Additional information is available at the end of the chapter

1. Introduction

Operations are the processes by which people, capital, and materials (inputs) are combined to produce the services and goods consumed by the public (outputs). Essentially, operations add value to the final product over and above the product's cost. Operations Management (OM) is the functional area of business primarily devoted to the creation, planning, and management of the resource capabilities used by a firm to create products or services. Galloway (2000) viewed operations management as all activities concerned with the deliberate transformation of appropriate range of resources to produce the organisation's intended outputs. This transformational model may be represented as in Fig 1. Inputs are the resources such as personnel, capital, equipment, information and technology while outputs include the actual delivery of required goods and services.

Figure 1. Transformational model of operations management

The resource capabilities comprise the inputs such as the work force (e.g., skills), technology (e.g., manufacturing equipment and information-based technology), and processes (e.g., supply chain, inventory-distribution system, quality control system, material flow system, production planning methods, monitoring system, etc.) all of which typically represent a significant portion of a firm's total costs and controllable assets. Since resource capabilities determine the types of products and services a firm can offer to the marketplace as well as the associated cost (price), quality attributes, and lead-times necessary to meet demand, the operations function is a critical driver of competitive advantage. Moreover, recent forces such as technology change and increased competition in cost, time and quality have elevated the distinctive competence that can be obtained from the effective management of operations function.

Healthcare operations management is considered as the quantitative management of supporting business systems and processes that transform resources (inputs) into health care services as outputs (Langerbeer II, 2008). Pharmacy operations are carried out within the healthcare system and have a mix of both intangible and tangible characteristics. Appropriate resources are transformed to create the pharmaceutical services which form intangible components of the operations. These services are knowledge - based and have high levels of customer interactions. The services accompany health commodities which are tangible products; the logistics and supply of which are major functions of operations management.

The objectives of the chapter are to

1.   Describe the scope of operations management in health care

2.   Justify the need for technology and automation in pharmacy operations

3.   Highlight some types of technology employed in pharmacy operations

4.   Examine human resource issues of operations and technology in the pharmacy

5.   Highlight process workflow of prescription filling in a pharmacy

6.   Describe process improvement approaches to optimise patient flow in a pharmacy.

2. The scope of operations management in healthcare

Operations management is the set of intrinsic or internal processes and decisions that help address costs, process, technology and productivity. There are obvious lapses in how the processes and systems of healthcare are managed hence the need for a greater focus on applying management science to improve the processes and outcomes. Quantitative management implies the use of analytical tools as well as extensive use of process and quality improvement techniques to drive improved results. Similarly, using technology to further automate and streamline some processes in healthcare operations can help reduce costs and maximise efficiencies. The scope and functions of operations management include strategies to reduce costs and variability, improving logistics flow, quality of customer service and productivity and continuously improving business processes.

Essentially, operations management of pharmacies is a discipline of management that integrates scientific or quantitative principles to determine the most efficient and optimal methods to support pharmaceutical services in patient care delivery. There must be an adoption of operations management techniques into pharmacy practice to help drive improvements and efficiencies. For instance, incorporating queuing theory and scheduling optimisation methods help to reduce wastes and cycle time in patient care. The scope of operations management of the pharmacy includes all functions related to management systems and business processes underlying the delivery of pharmaceutical services. This includes extensive focus on the process workflow, physical layout, capacity design, physical network optimisation, staffing functions, productivity management, supply chain and logistics, quality management and process engineering.

3. Technology and automation in pharmacy operations

Technology is an increasingly important element in operations management. It may be defined as "anything that replaces routine or repetitive tasks that were previously performed by people or which extends (or enhances) the capability of people to do their work" (Rough, 2001). Merriam-Webster (2009) defines technology as "the practical application of knowledge especially in a particular area" and "a capability given by the practical application of knowledge". On the other hand, automation refers to 'any technology, device or machine that is linked to or controlled by a computer and used to actually do work that was previously done by humans' (Rough, 2001). In essence, all automation is technology, but the reverse is not necessarily true.

Automation is a trend in technological development, which seeks to eliminate direct manual involvement in control procedures; whereas, mechanisation is a component part of automation and it is concerned with replacement of manual labour with machine (Encyclopaedia Britannica, 2010). The use of technology and automation is paramount in the focus of operations management on improvements of business support services. Technology as an enabling factor has an indirect impact on work and it should be considered whenever quality and efficiency is low, particularly in processes such as dispensing of medicines in a pharmacy. These activities are repetitive in nature but amenable to some basic technology. The decision to substitute technology for labour in such support services is the only way to reduce processing and transactional costs over the long run.

The Local Enterprise Authority (2009) identified three types of technology; these are product, process and support technologies. Product technology is embodied in the organisation's product and it is often an important element of the product. It provides the product's innovative features, improved performance, and the very materials that goods are made from. Scientific knowledge is applied in development and production of medicines by pharmaceutical manufacturers and other healthcare researchers for the purpose of improving health and wellbeing. For this reason, medicines and other pharmaceutical products can be regarded as technology. Examples are drug delivery systems such as transdermal patches and implantable computer incorporated drug administration devices. These are classified as product technologies covered in the area of pharmaceutical technology (Bozzette *et al.*, 2001).

Process technology is employed in the production process and refers to the actual method used to transform inputs into outputs or finished goods and services. It concerns the equipment used, the operations performed on materials or customers in the service systems and such technology could be manual, automatic or mental. In most cases, a range of technologies is available for defining a specific process and each one has advantages and disadvantages, which must be weighed in the light of a firm's competitive priorities. Technology is also often a crucial part of the process to produce and/or deliver the product, for example digital technology to produce and deliver published medicine information materials for use in patient care. On the other hand, support technologies are used to perform certain other activities that are not embodied in the product or production process of an organisation. Among such technologies are information and communication tools such as software packages, computer networks, and quality assessment technology.

The goal of operations management is effective and efficient use of resources which is enabled by technology; hence its need in the management of pharmacy operations. The benefits of technology to support pharmacy operations include improvement in safety, efficacy and economy of medicines (Slee *et al.*, 2002). The compelling need for technology and automation is partly the result of innovations in creation and design of new technologies and increased labour costs which call for more cost effective production systems and operations (Garsombke and Garsombke, 1989). Developments in technology can be seen as developments in the innovation process and technological innovation is the first commercial introduction of a new technology, which may take the form of a product, process or service (Ilori, 2006).

The use of technology in pharmacy operations improves medication safety in patient care and also enhances efficiency of the medical process. Furthermore, it improves the documentation of care (Kelly, 2006). Nevertheless, it is believed that the application of sophisticated technology in operations management poses its own challenge; while technology improves medication safety and processes in terms of reduced medication errors, there is little evidence to suggest that any single technology has the potential to radically reduce adverse events. However, there are incremental benefits to be derived from each and full benefits may only be realised when several technologies are used and integrated (Slee *et al.*, 2002).

4. Some types of technology employed in pharmacy operations

Various types of technology have found application in pharmacy operations. They can be grouped into mechanisation technologies, quality assessment technology, information and communications technologies (ICTs), automation technologies and the newly evolving biotechnology

4.1. Mechanisation technologies

Mechanisation has been described as the use of machines, either wholly or in part, to replace human or animal labour (Encyclopaedia Britannica, 2010). It serves as a means of providing human operators with machinery to assist them with the physical requirements of work. Example of mechanisation in pharmacy operations within the hospital is the use of electrical mixing vessel in compounding medicines. The technology for compounding medicines include facilities for making individualised doses of medicines such as intravenous feeding solution for patients unable to eat, or doses of anticancer chemotherapy.

Compounding in pharmacy operations can be described as the creation of a drug product by mixing ingredients (Jeffcoat, 2004). Sometimes the needs of a patient cannot be met by a licensed medicine as in the case of a child for whom a particular medicine is not commercially available as a liquid. In such circumstances, suitable products that are required to meet the same high standard as those of pharmaceutical industry can be prepared in the hospital pharmacy.

The tools and equipment used for compounding are simple laboratory equipments such as mortar and pestle, glassware and simple machines. Similarly, tablet counting for pre-packaging, repackaging and dispensing is one of the routine tasks in the hospital pharmacy. The traditional technology adopted has been the counting tray but the tablet counting machine (or tablet counter) is now available.

4.2. Quality assessment technology

Quality assessment technology is used for the task of assessing product quality. The technology used to assess the quality of drug products spans a wide spectrum of technological artifacts, some of which are chemical reagents, chemical and biochemical equipment and instruments as well as high precision instruments for pharmaceutical analysis. However the main thrust of quality assessment of drug products is chemical analysis.

Counterfeiting of pharmaceutical products has assumed a global concern and one of the measures for checking this menace is the use of low cost mobile quality assessment technology. The GPHF-Minilab® is a form of mobile quality assessment technology for detecting counterfeit and substandard quality medicines (Jahnke et al., 2001; Jähnke, 2004). The equipment is a portable, tropics-compatible and easy-to-use mini laboratory, developed mainly for use in developing countries, which are particularly affected by counterfeit drugs.

The GPHF-Minilab® is designed so that all the devices necessary for testing the drugs can be accommodated in two portable units, each about the size of a suitcase and the GPHF-Minilab® can be used without an external energy source. The mini-laboratory contains, in addition to the reference substances for the targeted active ingredients, all the necessary laboratory appliances such as test tubes, pipettes, pre-coated TLC plates and developing chambers as well as battery-powered UV lamps set to different wavelengths and the instructions and necessary solvents and reagents for the tests.

4.3. Information and communications technologies (ICTs)

Information and communications technology (ICT) has been described as consisting basically of information handling tools which include a varied set of goods, application and services that are used to produce, store, process, distribute and exchange information (Hamelink, 1997). It is an umbrella term that includes all technologies for the manipulation and communication of information such as radio, television, cellular phones, computer network and satellite systems as well as the various services and applications associated with them.

Hamelink (1997) described ICTs as those technologies that enable the handling of information and facilitate different forms of communication which include capturing technologies (e.g. camcorders), storage technologies (e.g. CD – ROMS), processing technologies (e.g. application software), communication technologies (e.g. Local Area Networks) and display technologies (e.g. computer monitors).

The overall ICT infrastructure comprises the computer and communication technologies and the shareable technical platforms and databases (Ross et al., 1996). Most of the technologies adopted in pharmacy operations are ICT-based technologies. The key ICT-based technologies to be discussed include the computer, other computer-based technologies and the telephone.

There has been earlier report that differentiated between telepharmacy and e-Pharmacy (Sood et al., 2008). Telepharmacy connotes the delivery of pharmaceutical to clients in a remote location. It is not a technology but a method which employs technology whereas e-Pharmacy is an innovative concept associated with electronic transactions, e-prescription systems, decision-support systems, among others, in the delivery of pharmaceutical services. In e-pharmacy, the pharmacist provides professional services to clients by electronic means, usually an ICT (e.g., telephone, internet, etc.). Peterson (2010) described telepharmacy as a branch of pharmacy practice that allows pharmacists to provide pharmaceutical care to patients at a distance through the use of the state-of-the-art telecommunications technology which allows a licensed pharmacist to supervise telepharmacy site through the use of video conferencing technology.

4.3.1. Computer applications in pharmacy

The computer is an electronic device that is capable of accepting data in a prescribed form, and processing and supplying the results in a specified format as information or signal to control automatically some machine or process (Graf, 1999). It is extremely fast, accurate and useful as an information processing device. In the contemporary world, the adoption of computers is integral to technological advances in any operational system.

The versatility of the computer for processing and communication of data has made it useful in pharmacy operations. The computer software is the adaptable part of the computer system but the proficiency of a user in handling the computer is a major factor in its usefulness. There is a minimum level of capability required for the successful use of a computer and this can influence the acceptance or otherwise by a potential user (Dixon and Dixon, 1994). The application could be home-developed or enterprise package and it could be stand-alone or integrated with other applications or technologies. According to Kling (1996), the relevant factors in the effect of technology, particularly computerisation at the workplace, pertain to gender equity, control, support systems, social design and new ways of organising work; but no Single logic has been applied towards changing the work system with computerisation. Computerisation has found immense application in pharmacy operations because of the benefits of enhanced efficiency and effectiveness of the work system. Staff efficiency becomes evident as more time is freed from routine tasks (Afolabi, 2005). Such available time may then be spent on patient counselling and other clinical functions to reduce potential medication errors for better health outcome. The computer system has been found to be the most widely adopted ICT in pharmacy operations in the hospitals (Perri *et al.*, 1988; Santell, 1995 and Quick *et al.*, 1997). The system can operate at the departmental level, at the hospital level or even networked with the world wide web. The computer-based technologies currently employed in pharmacy operations include information support applications, decision support applications, bar-code technologies, the personal digital assistants (PDAs) and other hand-held devices.

4.3.2. Computer-based technologies

4.3.2.1. Information support applications

Clinical informations systems are those that are integrated through the whole hospital system as hospital information system such as **electronic health record** or its variant, the **electronic medical record**, designed to capture the all clinical and administrative information about a patient and has come to be the main tool of any clinical information system. The record contains patient demographics, medical history, previous admission information, previous surgery information, and obstetric history. **Computerized provider order entry (CPOE)** refers to any system in which clinicians directly enter medication orders (and, increasingly, tests and procedures) into a computer system, which then transmits the order directly to the pharmacy (Grizzle *et al.*, 2007). **Electronic prescribing** or **e-prescribing** involves a process of using a computer to enter, modify, review and output or communicate prescriptions electronically to a patient's pharmacy (ASHP, 2007). It is used for sending an accurate, error-free and understandable prescription directly to a pharmacy from the point-of-care and further enhances the quality of care and patient safety by integrating medication order into the overall process of medical care delivery (Thompson and Brailer, 2006).

The pharmacy information system could operate within the hospital for managing inpatients or linked with other healthcare facilities such as community pharmacies for outpatients. An example is the Pharmacy Care Plan which is a computer based form used for keeping patients record for the purpose of pharmaceutical care (Winslade *et al.*, 1996). It integrates data from various other systems within the hospital to improve therapy, safety or costs.

4.3.2.2. Decision support applications in pharmacy operations

Decision support applications can be classified into clinical decision support systems (CDSS) and Clinical Pharmacokinetics Computer Programs (CPCP).

Clinical decision support systems (CDSS or CDS) are interactive computer programs, which are designed to assist physicians and other health professionals with decision making tasks. They provide clinicians with patient-specific assessments or recommendations to aid clinical decision making (Kawamoto *et al.*, 2005). The methodology of using CDSS forces the clinician to interact with the CDSS utilising both the clinician's knowledge and the CDSS to make a better analysis of the patient's data than either human or CDSS could make on their own.

Clinical pharmacokinetics computer programs are computer software for use by pharmacists in analysing serum concentration data. They are employed for patients to individualise dosage regimens of highly toxic medicines and those with narrow therapeutic windows.

4.3.3. Bar-code technologies

A bar-code is an optical machine-readable representation of data (Okon, 2005). It is usually a series of vertical bars and spaces. The bars and spaces represent digitally encoded information which is designated by the width of each bar and space. The height of a bar code is irrelevant although it must be sufficient in size to allow an operator to easily read each bar with a scanning device (Okon, 2005). Since the codes are digitally made they can only be read and interpreted by means of electronic machines designed for this purpose hence bar-codes are normally read by optical scanners called bar-code readers, or scanned from an image by special software (Okon, 2005).

Bar-code technologies are applied in two ways either as Bar-code-assisted dispensing (Bar-D) or bar-code at the point of medication administration system (BCMA). They are designed to prevent medication errors in hospitals (Sakowski, 2005). They work with hardware that consists of a handheld scanning device connected to a wireless mobile computer. This mobile hardware communicates with other information equipment in the system through a server to record transactions. Some systems utilise a bar-code scanner-enabled personal digital assistant (PDA) to scan, confirm and store medication transaction information.

4.3.4. Personal Digital Assistants (PDAs) and other hand-held devices

These are small mobile computers such as the blackberry, apple iPhone and "smart phones" (Anderson et al., 2010). They usually have integrated ICT applications such as telephone, camera, scanner, etc. Their ability to handle vast amount of information and instantaneous speed of operation including switching on and off makes them vital for medicine information services.

4.3.4.1. The use of telephone in pharmacy operations

The Short Message Service (SMS) on mobile phones, as a form of e-messaging has been employed in promoting adherence to medicine therapy especially in the treatment of chronic diseases such as hypertension, HIV/AIDS, diabetes and osteoarthritis. Adherence is the degree to which patients conform to a given treatment plan or the extent to which patients take their medications as prescribed by healthcare providers. A review of published studies reported that up to 10% of hospital admissions are related to non-adherence.

One of the important aspects of pharmaceutical care is the provision of medicine information service (MIS) to patients. Mobile telephones are readily available and this form of technology can be used to communicate medicine information to patients with chronic ailments.

4.4. Automation technologies in pharmacy operations

A number of technologies have been used to automate various aspects of pharmacy operations. Their aim is usually to replace manual labour and promote safety and economy. The four main benefits that result from the use of automation technologies in pharmacy operations include reduction in errors (Slee *et al.*, 2002), improved use of space and workflow (Cairns, 2002) improved stock control and stock management (Martin, 2002) and improved departmental efficiency (Fitzpatrick *et al.*, 2005) which reduces patient waiting times and allows pharmacists to use their time more effectively in pharmaceutical care.

Basically, there are three types of automated devices namely automated compounding and/or counting devices, automated medication supply systems and automated checking devices (Slee *et al.*, 2002; Fitzpatrick *et al.*, 2005). They could be for inpatient or outpatient services in the hospital (Anderson *et al.*, 2010).

4.4.1. In-patient pharmacy automation technologies

The available inpatient pharmacy automation technologies include the automated checking device, smart infusion pumps, automated dispensing cabinets (ADCs) and robots (Anderson *et al.*, 2010).

The automated checking device is a fully automated device which confirms, after a drug is prepared for distribution but prior to delivery to the patient, that the right drug has been labelled correctly for the right patient using technology such as bar-coding (Fitzpatrick *et al.*, 2005).

Smart pumps

The "smart" infusion pump is a medication administration equipment for providing medication infusion at a specified rate. It has a computer "brain" with a database of standard intravenous (IV) preparations from which the required item can be selected. It can trigger alert when any of the limits set for it concerning dosage or other data has been exceeded.

Automated dispensing cabinets

Automated dispensing cabinets (ADCs) are computerised packaging equipment in which a consignment of medicines in unit dose packaging (as in a ward stock) are contained in locked drawers. The machine is able to dispense medications electronically in a controlled fashion and track medication use. The storing and picking of products and the labelling stage of the dispensing process are automated (Slee et al., 2007). ADCs dispense the required medicines and record the identity of the user, the patient and the medicines dispensed.

Robots

The pharmacy robots add kinetic components that mimic human activity to the capability of the ADCs by way of preparing, dispensing, and distributing medicines to various locations in the hospital. It employs the bar-coding technology for identifications by labelling or scanning as necessary.

4.4.2. Outpatient pharmacy automation technologies

4.4.2.1. Automated drug dispensing machines

Automated drug dispensing machine (ADDM) also called automated compounding or counting device is an automated device that compounds, measures, counts, and/or packages a specified quantity of dosage units of a designated drug product (Fitzpatrick et al., 2005). The ADDM is interfaced with the pharmacy information system and it can fill, label and deliver prescriptions received from the system (Anderson *et al.*, 2010).

4.4.2.2. Automated kiosks

Automated kiosks are devices that operate like Automated Teller Machines holding the medication that has been filled by the pharmacist and delivering the medicines to patients at their convenient time (Anderson et al., 2010). They allow for payment at the point of receipt.

4.5. Biotechnologies in pharmacy operations

The study of genome, the entirety of an organism's hereditary information, has led to the discovery of the sequence of genes in the human DNA (Ridley, 2006; Anderson *et al.*, 2010). The DNA is the means by which genetic traits are transferred and the structure of each organism's DNA sequence determines the genotype and hence the characteristics of the organism. This knowledge has a great potential for applications in producing medicines that are tailored to each person's unique organic characteristics as well as to selectively target pathogens in disease management. The full benefit of biotechnology in pharmacy practice is still potential although its magnitude probably cannot be imagined for now.

5. Human resource issues, technology and automation of work

The workforce of an organisation, though viewed as part of the resources, actually forms the interface between the customers and the organisation and is significant in the management of operations. In fact, the workforce represents the organisation to the customer and is responsible for the system to be functional. It is necessary therefore to temper the technical focus of traditional operations management with the reality of human behaviour and employee satisfaction (New, 1998).

Automation as a trend in technological development is an enabling factor in the work system and should be considered in quality improvement strategies of operations management. The direct impact of automation could be felt at the micro level of a work system as this technological change can alter the nature of the tasks, work cycles, skill requirements, and worker interactions. With increasing automation, jobs tend to become more demanding, varied, interesting and challenging for many workers; although in some cases, such changes may be of a temporary nature. Technical know-how tends to become more important and workers may expect increased job content together with increased demands on skills, knowledge and training.

Automated systems may result in greater complexity and responsibility, and therefore greater intrinsic rewards, but often at the expense of worker inactivity. Furthermore, automation has considerable effects on social interactions. The greater distance between workers caused by automated systems may result in reduced social relationship, and also, there is an increased separation of workers from both operations and their outputs. On the other hand, in automation there could be an increased contact between workers and their supervisors leading to improved worker - supervisor relations. Similarly, increased training needs are often associated with the wider responsibilities of automated jobs.

Occasionally, increased stress may be experienced in work situations approaching full automation. This may be as a result of emphasis on vigilance and monitoring duties. The importance of minimising process disruption, the consequences of breakdowns and comparative inactivity of workers also contribute to stress conditions in automation.

One reason for the automation of work is the desire to remove workers from unsatisfactory working environments. Such criteria for automation include risk of accident, monotony of work, excessive physical stress and bad working conditions (Wild, 1999). It may be necessary, therefore, to have some means to check the acceptability or appropriateness of working conditions. This will enable existing work systems to be appraised in order to establish whether there is the need for a change and similarly a new work system design can be checked for acceptability or appropriateness by the workers before implementation.

Although automated systems can contribute to an organisation's complexity, they can also help the organisation to cope with such complexity. The speed and accuracy of the computer, as an automated technique, seem attractive for decentralised decision making while top management has time for planning and innovation. Computer - based information could improve the managers' capabilities making them better employees and more important to the success of the organisation. Some routine, clerical activities are easily adaptable to computerisation, and procedures may be developed to handle structured repetitive tasks. For instance, in a pharmacy, such repetitive tasks as prescription billing, documentation, and stock control functions may be taken over by the computer (Slee *et al.*, 2002). In these situations, the work is repetitive and can be described in a clear procedure for routine transaction processing. With this development much of the mundane, tedious routine jobs of prescription processing may be automated while the unstructured, non routine and more skilled decisions can then be handled by the pharmacists.

In clinical functions, creativity, insight and professional judgment are needed on the part of the pharmacist who has to advice on choice of medication in a patient's therapeutic plan. Much of the problem solving activities in these areas require knowledge of the medicines and their therapeutic efficacy. Assistance from computer - based information systems helps management decision making and this could improve a pharmacist's advisory skill in patient care. Thus, rather than replace or displace pharmacists' ranks, introduction of automated techniques, such as automated medicine information systems, should create a demand for more highly qualified and better-trained pharmacists. Computer training and enhanced clinical skills should therefore be an integral part of pharmacy education and retraining programs. The workers in an organisation should be treated as a major long term investment with enhanced skills in order that the company may realise its full potentials and in such a way that these employees may feel appreciated (Scaborough and Zimmerer, 2000).

Much as technology is desirable to enhance operational effectiveness, the implementation of new technology will benefit from due attention to factors connected to employee perceptions and acceptance of the change. Genus and Kaplani (2000) examined the need to consider the behaviour of employees in connection with operations and changes in the design of the work system brought about by technology. He concluded that shared values between an organisation and the work force would facilitate the implementation of operational innovations. Similarly, Savery and Choy (1999) opined that a strategic choice of technology should include due attention to human resource management issues and a consideration of the assets and appropriate facilities for operational effectiveness. Therefore, for an organization to realise its full potentials, the workforce should be motivated with a conducive and friendly work environment.

It has been shown that content and design of the work system can also be a source of motivation or otherwise for workers (Scaborough and Zimmerer, 2000). A lack of employee motivation and the absence of shared values could pose as barriers to employee involvement in and commitment to continuous improvement objectives in an organisation (Afolabi and Oyebisi, 2007a). Appropriate staff training and a high degree of motivation may achieve the harmony of staff perception and organisational intention in the adoption of technology to enhance performance.

The pharmacy unit is a service operating system and it is expected to provide relevant and adequate infrastructure and tools required to satisfy client needs in terms of appropriate medicine supply and information services as well as facilitate social and appropriate professional interactions in the health system. In addition, the system should enhance adequate patient flow.

6. Operations management and patient flow in hospital pharmacies

Patient flow optimisation opportunities occur in many healthcare settings especially at the pharmacy units of hospitals in resource limited countries; where patients wait in lines to fill their prescriptions. There could be many causes of poor patient flow in such circumstances but the major cause is variability of scheduled demand. Variability is the inconsistency or dispersion of inputs and outputs and this threatens processes because it results in uncertainty. For instance, if there are 50 patients typically seeking care at the outpatient pharmacy unit within a certain time period and 100 appear the following period, it becomes difficult for staff to control waiting times and to manage patient flows. Improving flow means seeking higher throughput or yields for the same level of resource input.

Throughput is the rate or velocity at which services are performed or goods are delivered. If patient volumes double but a hospital maintains the same historical inventory levels of pharmaceutical supplies, this represents significant improvements in material flow, because assets have higher utilisation and turns. Staffing and resource consumption should be tied directly with patient volumes and workload; if patient volume increases, the resources should also increase.

Operations analysis helps to track demand variability which is consequent on a surge in patient demand. Managing this variability allows a change in staffing mix and scheduling to accommodate the changes in demand. Staffing at the peaks will cause excessive costs while staffing for the valleys or low points will cause long lines periodically due to limited resources and therefore service quality issues; on the other hand, staffing for the average demand is the most common but suboptimal approach. Managing patient flow is an operating issue in a healthcare organisation but effective and timely management provides an opportunity for desired outcome. Significant opportunities exist to improve capacity and reduce costs by improving patient flow using the formal methods of process improvement.

7. Process improvement approaches to optimise patient flow in a hospital pharmacy

Filling of prescriptions forms the core of pharmacists' activities in the hospitals in Nigeria. However, considerable delays were observed in dispensing operations in the pharmacies and these delays result in long patient queues, particularly at peak periods. A process improvement project was carried out to examine dispensing procedure at the pharmacy unit of OAUTHC, and to identify aspects of the process design (task elements) that contributed to patient delays.

Data were collected by direct observation of dispensing workflow and a time study of the procedural elements. A checklist was used for a systematic recording of the various activities and queuing models were used to characterise the waiting line structures at the pharmacies with a view to simulating optimal utilisation factors (Pw) for the service channels. Time elements of dispensing operations revealed extensive delay components which contributed to excessive patient waiting time.

The operational problems identified in the work process were due to tortuous procedure for prescription filling and the volume of transactions at the cashier stand. The operating characteristics of the paying systems were not optimal with only one service channel. The delays observed were substantially due to existing work procedure in the pharmacy and the volume of manual transactions which are amenable to newer technologies. Some of the dispensing task components were not essential and these elements could be removed or merged in order to reduce patient waiting time (Table 2). The details of the project are presented in section 7.2.

7.1. Process maps

Extended waiting time to fill prescriptions in hospital pharmacies can be addressed by using some process, improvement tools. A process map or flowchart is a graphic depiction of a process showing inputs, outputs and steps in the process. Figure 2 shows a process map that illustrates the activities involved in prescription filling at the pharmacy unit of a university teaching hospital in Nigeria. Figures 3a and 3b present the staff process chart, indicating the tasks and delay components of the operating system.

Typically, process maps are used to understand and optimise a process and the process may be charted from the viewpoint of the material, information, the worker carrying out the work or the customer being processed. In Figure 2, the customer being processed is the prescription-carrying patient within the healthcare. Process mapping is a basic quality tool and an integral part of most improvement initiatives. The steps for creating the process map include an observation of the process and a description to ensure the real activities are captured. Boundaries are determined for the activities and the process tasks and subtasks are listed and arranged in order as a written procedure or protocol (Table 1). A formal flow chart is then generated using standard symbols for process mapping. This is then checked for accuracy and additional data on process performance may be added depending on the purpose of the flowchart. Service blueprinting is another quality improvement tool and a special form of process mapping as shown in figure 4. The process is mapped from the point of view of the customer.

Basically, the purpose of a service blueprint is to identify points where the service might fail to satisfy the customer and then redesign or add controls to the system to reduce or eliminate the possibility of failure. Three distinct actions are delineated in serve blueprinting and these are the customer actions which show the nature of customer involvement in the process and the interactions, onstage actions which are visible to the customer, the backstage and support processes which are not visible to the customer. A service blue print specifies the line of interaction, where the customer and service provider come together, and the line of visibility, that is, what the customer sees or experiences, the tangible evidence that influences perceptions of the quality of service.

Process maps provide a visual representation of the process and this offers an opportunity for improvement through inspection. The maps allow for branching in a process and provide the ability to assign and measure the resources in each task. Process mapping are the basis for modelling the process via computer simulation software or using other process technology. Simulation is a modelling technique that may be used to evaluate the effects of possible changes on an operating system. The process of simulation consists of model development, model validation and an analysis of the output to optimise a process or manage risk. Discrete event simulation using the queuing theory may be used to model system flows as an improvement strategy. The following section (Section 7.2) applies these concepts to process improvement in a hospital pharmacy with emphasis on patient waiting time.

7.2. Operations management project : The case of OAUTHC pharmacy

In order to demonstrate possible application of the process improvement tools described previously, a patient- flow process improvement project at the pharmacy unit of a teaching hospital was examined.

It was observed that outpatients experience considerable delays to fill prescriptions at the pharmacy unit of a university teaching hospital (OAUTHC) and this was identified as an important area on which to focus improvement efforts. The goal of the project was to reduce patient waiting time while optimising capacity utilisation of the pharmacy resources in the hospital.

The first step for the project team was to identify the need to facilitate patient flow with improved dispensing operations at the pharmacy. The specific tasks in the project were:

1. Observe dispensing operations and write out the detailed steps using a checklist (Table 1)

2. Develop a flow chart of dispensing to outpatients (Figure 2)

3. Measure specific metrics for the procedural elements (Table 1) taking an average of observations for one week

4. With the flow chart and specific metrics, identify the delay components of the operating system(Figure 3 &Table 2)

5. Use simple process improvement techniques to make changes in the process, then measure the results

6. Collect data needed to build a realistic simulation model based on queuing theory (Tables 3 & 4)

7. Develop the simulation model and validate it against real data

8. Use the simulation model to conduct virtual experiments to improve patient flow.

9. Implement promising improvements and measure the results of the changes.

Procedural elements	Time (min.)

A. Initial patient contact and information collection

1. Clerk receives prescription and takes prescription to the pharmacist for pricing and processing

2. Clerk gives the prescription to patient

Procedural elements	Time (min.)

B. Payment time
1. Patient arrives at the queue
2. Patient gives prescription to the accounts clerk
3. Patient receives the prescription from the clerk and pays for medication

C. Submission of prescription and dispensing
1. Clerk receives prescription from the patient after payment and takes the
prescription to the pharmacist for processing
2. Pharmacist receives prescription from the clerk and edits the form
Pharmacist telephones physician for refill authorisation
Pharmacist queries any inappropriate prescribing

D. Obtain and package medication
1. Pharmacist writes and affixes label to the container
Attendant brings the medication and instructs patient
Pharmacist counsels patient (occasionally)
Pharmacist gives medication

E. Final clerical processing and cleaning
1. Pharmacist files patient prescription
2. Attendant performs clean- up

Table 1. Procedural elements of outpatient dispensing in the pharmacy

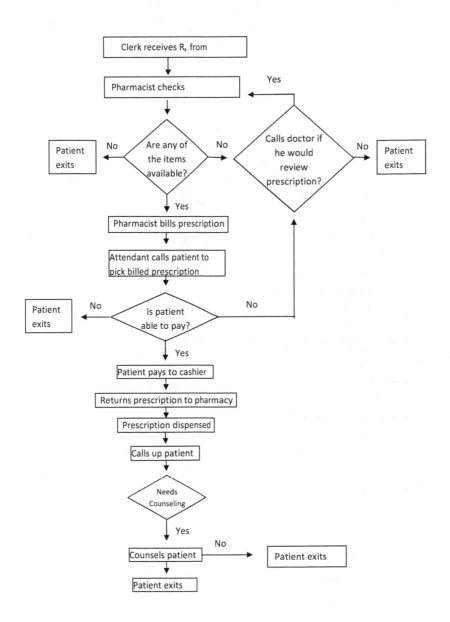

Figure 2. Process flow chart for prescription filling to outpatients

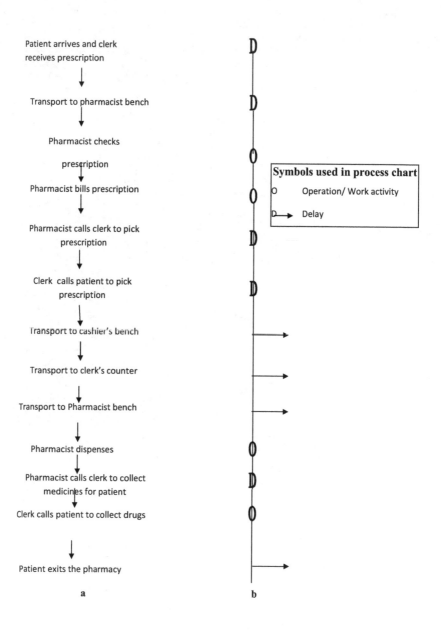

Figure 3. a: Staff process chart for dispensing to outpatients; b: Delay and process chart for dispensing to outpatients

Components of dispensing procedure	Time spent (min.)			
	Processing		Delay	
	Mean	%	Mean	%
Patient waits for clerk to collect prescription	*	*	0.76	4.45
Clerk takes prescription for billing	0.15	0.88	*	*
Prescription placed on dispensing table for billing	*	*	0.44	2.57
Pharmacist bills and reviews prescription	0.62	3.63	*	*
Billed prescription placed on dispensing table for clerk	*	*	0.60	3.51
Clerk takes prescription to patient	0.18	1.05	*	*
Patient takes prescription to cashier	0.14	0.82	*	*
Patient waits on queue for payment	*	*	8.68	50.79
Cashier collects money, issues receipt and records	1.42	8.31	*	*
Patient collects receipt and prescription; and takes them to the clerk	0.18	1.05	*	*
Patient waits for the clerk to collect receipt and prescription	*	*	0.56	3.28
Clerk takes prescription and receipt to pharmacist	0.12	0.70	*	*
Prescription and receipt placed on dispensing table for pharmacist to fill	*	*	0.58	3.39
Pharmacist fills prescription	1.38	8.07	*	*
Medicine placed on dispensing table for clerk	*	*	0.86	5.03
Clerk takes medicine to patient, instructs and dispenses	0.42	2.46	*	*
Total	4.29	25.10	12.80	74.90

Total patient waiting time = 17.09 min. * = Nil

Table 2. Observed time distribution between processing and delay components for each task of the dispensing process at the pharmacy.

Hospital	Waiting line parameters								
	No of Channels	λ/min	μ/min	Po	n_s	n_q	t_s(min)	t_q(min)	P_w
	4	0.62	0.40	0.20	1.60	0.05	2.58	0.08	0.08
OAUTHC	3	0.62	0.40	0.20	1.82	0.27	2.94	0.44	0.25
	2	0.62	0.40	0.12	3.83	2.28	6.18	3.68	0.66

Where:

λ = Arrival rate

μ = Service rate

P_o = The probability of no client in the queue system

n_s = The average number of clients in the queue system

n_q = The average number of clients in the queue waiting for service

t_s = The average time a client spends in the queue system (waiting time + service time)

t_q = The average time a client spends in the queue waiting for service

P_w = The probability that an arriving client has to wait for service

(utilisation factor or capacity utilisation of the facility)

Table 3. Operating characteristics of patient waiting lines in the pharmacy

Hospital	Waiting line parameters								
	λ/min.	Models	μ/min.	n_s	n_q	t_s(min)	t_q(min)	P_w	P_o
	0.60	A	0.67	8.6	7.68	14.30	12.80	0.90	0.10
OAUTHC	0.60	B	0.86	2.31	1.64	3.85	2.73	0.70	0.30
	0.60	C	0.92	1.88	1.24	3.13	2.07	0.65	0.35

Where:

λ = Arrival rate

μ = Service rate

P_o = The probability of no client in the queue system

n_s = The average number of clients in the queue system

n_q = The average number of clients in the queue waiting for service

t_s = The average time a client spends in the queue system (waiting time + service time)

t_q = The average time a client spends in the queue waiting for service

P_w = The probability that an arriving client has to wait for service

(utilisation factor or capacity utilisation of the facility)

Table 4. Operating characteristics of one cashier service channel with varied service rates

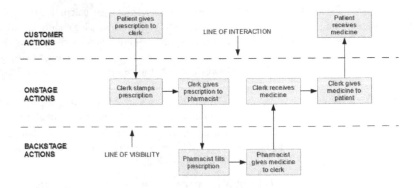

Figure 4. Service blueprint of prescription filling in a hospital pharmacy

7.3. Operating characteristics of outpatient waiting lines at OAUTHC pharmacy using the simulation model

A patient's experience of waiting can radically influence his/her perceptions of service quality and various studies have been carried out on queues in pharmacy systems and strategies to promote customer satisfaction (Lin *et al.,1996;* Lin *et al., 1999).* Simulation studies and queuing theory had been used to explore how changes in patient arrival rates and the time taken to dispense prescriptions were likely to affect patient waiting times (Toohey *et al., 1982,* Afolabi and Erhun, 2003). Some of the authors used simulation model to demonstrate that waiting times could be reduced by staggering pharmacists' break periods. Simulated analysis of patient queues in other studies examined the relationship between staff organisation, patient arrivals and waiting times in hospital pharmacies (Slowiak *et al.,* 2008).

Queuing models give descriptions of queuing systems and such descriptions provide valuable inputs to the decision-making process of an operating system. For instance, given a queuing model, the descriptions of a queuing system that may be obtained include the probability that there will be a number of clients waiting in the queue and the probability of the systems being idle. Similarly the description gives an idea of the average number of clients waiting in the queue and the expected number of clients in the system (i.e. queue plus service facilities). The time estimates may also be obtained from the description as the model gives the probability of the service times, the average time a client spends on the waiting line and the total time required to get through the entire queuing system (Panthong and Daosodsai 2005; Vemuri, 1984).

The operating characteristics of patient waiting lines in the hospital pharmacy of OAUTHC using four service channels are presented in Table 3. The results indicate that patients queued for about 3- 5 seconds (t_q) and stayed for about 2 – 6 minutes in the system. This implies that with four dispensing pharmacists actively working, patients did not need to spend a long time in the pharmacies. However, personal observation showed that most of the time only two or three of the pharmacists attended to patients on the queues while the others were occupied with other activities in the pharmacies. The queuing characteristics were calculated using three or two servers and the results are presented in Table 3. With two service channels, patients spent about 6 minutes in the pharmacy while the time on the queues was up to 4 minutes. With four service channels, the mean utilisation factors (P_w - 0.04 – 0.08) of the facilities were quite low; this translates to a capacity utilization of 4% to 8%. On the other hand, the mean utilisation factors were highest (P_w - 0.08 – 0.66) for two service channels, that is, capacity utilisation of 13% to 66%. With these utilisation factors, the number of patients on the queues was still reasonable (1 – 2 patients). The number of dispensing pharmacists could be kept at two in the outpatient units while other pharmacists are engaged in other tasks in the pharmacies. However, the dispensing duties may be rotational to avoid undue stress as a result of being stationed at routine activities for the full work hours each day.

Similarly, the operating characteristics of waiting lines at the cashiers' counters are presented in Table 4. The values recorded for the parameters as shown in row A of the table were calculated from the observed mean arrival rates (λ) and mean service rates (μ). The results in row A for the hospital pharmacy indicate that patients queued for about 8 minutes before service while they spent more than 14 minutes in the paying system. This delay may be of concern considering the state of ill-health of some of these patients and the fact that they would have to join queues in some other service units of the hospitals. The mean utilisation factor of the operation is quite high ($P_w = 0.90$; capacity utilisation of up to 90%) but the resultant long queues (up to 9 patients waiting on the queue) and the waiting time may require additional service channels. Similarly, the length of queue and the waiting time can be reduced either by increasing the service rates or making a creative design change such as modifying the service channels to improve the process.

On the other hand, using the simulation technique, rows B and C of Table 4 present the queuing characteristics assuming utilisation factors (P_w) of 0.70 and 0.65 that is, capacity utilisation of 70% and 65% for the queuing systems. This analysis assumed a constant arrival rate at each facility but varying the service rates to obtain utilisation factors (P_w) of 0.70 and 0.65 gave the operating characteristics in rows B and C of Table 4. As the service rates increased, the performance measures of the waiting lines also improved remarkably and the number of people waiting in line was about two in the pharmacy. The probability values of no patient in the systems (P_o) also increased with improved service rates.

This result agrees with the findings of a previous study which evaluated the characteristics of waiting lines in some food service systems in Nigeria (Oyebisi et al., 1999). Service rates may be increased at the payment counters in order to reduce the total patient waiting time in the pharmacy. The application of modern technology in cash transactions and accounting records may help to speed the process; for instance, the adoption of new technologies such as electronic payment and computerised recording system may relieve the delay caused by manual operations and this can facilitate a smooth flow of the dispensing process.

The results of a previous study showed that the hospital pharmacists were well disposed to the introduction of such technologies to improve some aspects of service operations in the pharmacies (Afolabi and Oyebisi, 2007b). On the other hand, there may be the need for a process redesign or adding more parallel service channels. This latter option will increase the number of serving personnel with the attendant increased service cost in terms of physical arrangement and personnel costs. However, it is necessary to maintain an economic balance between reasonable utilisation factors and moderate sized queues in the operating system. Essentially, new systems should be carefully assessed to be sure they contribute more to the ultimate focus of enhancing pharmacist involvement in patient care rather than just speed the process of dispensing with little time for patient counselling. Pharmacists would definitely be more satisfied at their jobs if they were able to dispense what they consider to be a 'comfortable' number of prescriptions per hour that would enable them to counsel patients properly and attend to other clinical practice contributions to patient care.

Other strategies that could significantly speed the process of service delivery in the pharmacy include the adoption of new technologies such as electronic payment and computerisation of some of the pharmacists' activities. However, the use of new technologies may not necessarily contribute to better dispensing procedure and rational use of medicines. Other areas of positive turn around in service delivery could be a job redesign and continuing reorientation of pharmacists to contemporary and emergent techniques in clinical skills and medicines management.

8. Conclusion

The healthcare system continues to experience dynamic change as a result of factors such as government influence, economic pressures, the biotechnology development of medicines and the continued development of robotics and automation. The pharmacy is integral to the operations of health care delivery and it should grow from a medicine distribution system to a very effective means by which patient health care may be improved through proper interpretation of the prescriptions, monitoring and follow-up of patient medicine therapy regimens. The future of pharmacy profession, as part of the healthcare system, does not lie in the dispensing or distribution of medications but in the provision of information and medicine therapy recommendations to other healthcare providers and the patients concerning rational therapeutics. As a result of these advances, the pharmacist must be prepared to meet the needs of the healthcare system and most importantly, to adopt appropriate technology in the provision of pharmaceutical services.

Author details

Margaret O. Afolabi and Omoniyi Joseph Ola-Olorun

*Address all correspondence to: bunmafol@oauife.edu.ng, bunmafol@gmail.com

Department of Clinical Pharmacy and Pharmacy Administration, Faculty of Pharmacy, Obafemi Awolowo University, Ile-Ife, Nigeria

References

[1] Afolabi, M. O. (2005). "Evaluation of Operations Management of Pharmacies in Se-
     lected Teaching hospitals in Southwestern Nigeria", Ph.D. thesis (Unpublished). Ob-
     afemi Awolowo University, Ile-Ife, Nigeria, pp: 205-211.

[2] Afolabi, M.O. and Erhun, W.O. (2003): Patients' response to waiting time in an out-
     patient pharmacy in Nigeria. *Tropical Journal of Pharmaceutical Research* 2 (2):207-214.

[3] Afolabi, M. O. and Oyebisi, T. O. (2007a). Pharmacists' Perceptions of Barriers to Au-
     tomation in Selected Hospital Pharmacies in Nigeria. *Journal of Pharmacy Practice*, 20:
     64-71.

[4] Afolabi, M.O. and Oyebisi, T.O. (2007b). Pharmacists, Attitude to the Introduction of
     Automated Techniques in the Delivery of Pharmaceutical Services in Selected Niger-
     ian Teaching Hospitals. Journal of Pharmacy Practice 20 (1), 72-81.

[5] Anderson, P. O., McGuinness, S. M. and Bourne, P. E. (2010). Pharmacy Informatics.
     New York: CRC Press.

[6] ASHP (2007). American Society of Health-System Pharmacists' statement on the
     pharmacist's role in informatics. *Am J. Health-Sys. Pharm.*, 64: 200-203.

[7] Bozzette, S. A., D'Amato, R. D., Morton, S., Harris, A., Meili, R. and Taylor, R. (2001).
     Pharmaceutical Technology Assessment for Managed Care: Current Practice and
     Suggestions for Improvement. [Website]. Available from: <http://www.questia.com>
     [Accessed: 20 March, 2009].

[8] Cairns, C. (2002). Robots and Automation: a UK perspective. *Hospital Prescriber*, Eu-
     rope Spring 02. [Website]. Available from: <http://www.hospital prescriber.com>
     [Accessed: 8 September, 2010].

[9] Dixon, D. R. and Dixon, B. J. (1994). Adoption of information technology enabled in-
     novations by primary care physicians: model and questionnaire development. *Proc.
     Annu. Symp. Comput. Appl. Med. Care*, 1994, pp. 631–635.

[10] Encyclopaedia Britannica (2010). Mechanisation. Encyclopaedia Britannica Online.
     [Website]. Available from: <http://www.britannica.com> [Accessed: 7 April, 2010].

[11] Fitzpatrick, R., Cooke, P., Southall, C., Kauldhar, K. and Waters, P. (2005). Evaluation
     of an automated dispensing system in a hospital pharmacy dispensary. *The Pharma-
     ceutical Journal*, 274: 763-765.

[12] Galloway, L. (2000). Principles of Operations Management 2nd ed. Ohio: Int. Thomp-
     son Bus. Press.

[13] Garsombke, T.W. and Garsombke, D.J. (1989). Strategic implications facing small
     manufacturers: the linkage between robotization, computerization, automation and
     performance. *Journal of Small Business Management*, 27

[14] Genus A, Kaplani M.(2002) Managing operations with people and technology. *Intern Journ of Techn Manag.*; 23(1-3):189-200

[15] Grizzle, A.J., Mahmood, M.H., Ko, Y., Murphy, J.E., Armstrong, E.P., Skrepnek, G.H. *et al.* (2007). Reasons provided by prescribers when overriding drug-drug interaction alerts. *Am J Manag Care*, 13: 573-578.

[16] Hemelink, C. J. (1997). New Information and Communication Technologies, Social Development and cultural change. United Nations Research Institute for Social Development (UNRISD). UNRISD Discussion. No. 86.

[17] Ilori, M. O. (2006). "From Science to Technology and Innovation management." An Inaugural Lecture Delivered at Oduduwa Hall, Obafemi Awolowo University, Ile-Ife. May 30. Inaugural Lecture Series 191. O.A.U. Press Ltd., Ile-Ife. Nigeria.

[18] Jähnke, R. W. O. (2004). Counterfeit Medicines and the GPHF-Minilab for rapid drug quality verification. *PharmazeutischeIndustrie*, 66 (10): 1187-1193.

[19] Jähnke, R. W. O., Kusters, G. and Fleischer, K. (2001). Low-cost quality assurance of medicines using the GPHF-Minilab®. *Drug Information Journal*, 35: 941 – 945.

[20] Jeffcoat, M. K. (2004) Eye of newt, toe of frog. *J. Am. Dent. Assoc.*, 135 (5): 546-547.

[21] Kawamoto, K., Houlihan, C. A., Balas, E. A. and Lobach, D. F. (2005). Improving clinical practice using clinical decision support systems: a systematic review of trials to identify features critical to success. *BMJ.*, 330(7494): 765

[22] Kelly, W. N. (2006). Pharmacy, what it is and how it works 2nd edition. Boca Raton: CRC Press pharmacy education series.

[23] Kling, R. (1996). Computerisation at Work in Computerisation and Controversy; Value Conflicts and Social Choices (2nd edition). San Diego, CA: Academic Press.

[24] Langabeer II, J.R. (2008). Health care operations management: a quantitative approach to business and logistics. Sudbury,MA, Jones and Barlett Pub., pp 6-7.

[25] Lin, A.C, Jang, R and Sedani D. (1996): Re-engineering a pharmacy to facilitate patient counselling. *American Journal of Health-System Pharmacy* 53: 1558-64.

[26] Lin, A.C., Jang, R., Lobas, N., Heaton, P., Ivey, M. and Nam, B. (1999): Identification of Factors Leading to Excessive Waiting Times in an Ambulatory Pharmacy. *Hospital Pharmacy* 34 (6):707-712.

[27] Local Enterprise Authority. (2009). Technology Adaptation and Adoption. [Website]. Available from: <http://www.lea.co.bw/article.php?id_mnu=48> [Accessed: 9 May, 2009].

[28] New, C (1998)'The state of operations management in the UK – a personal view', *International Journal of Operations and Production Management 18(7-8):675-679*

[29] Oyebisi, T.O., Ilori, M.O. and Oyeleke, T.O. (1999): Characteristics of the waiting line in selected traditional food service systems in Osun State of Nigeria. Nigeria Institute

of Industrial Engineers (NIIE) Proceedings of 1999 Annual Conference held at Ibadan, p 74-88.

[30] Panthong S, and Daosodsai P. Work Analysis Model of Hospital Pharmacy Services at Banphai Hospital Thailand. Malaysian Journ Pharm Sci .2005; 3(2) 47

[31] Ridley, M. (2006). Genome. New York, NY: Harper Perennial.

[32] Savery, LK and Choy RM (1999) "Cutting edge' technology: an avant garde solution of operational effectiveness or careerist nightmare?", *Intern Journ of Techn Manag;* 17(1-2):145-156

[33] Slowiak, J. M, Huitema B. E. and Dickinson A (2008) . Reduce wait time in a hospital pharmacy to promote customer satisfaction. *Quality Management in Health Care.;* 17 (12): 112-127.

[34] Sood, S. P., Prakash, N., Agrawal, R. K. and Foolchand, A. A. B. (2008). Telepharmacy and e-Pharmacy: Siamese or discrete? *International Journal of Healthcare Technology and Management,* 9 (5-6): 485-494.

[35] Thompson, T. G. and Brailer, D. J. (2006). The Decade of Health Information Technology: Delivering Consumer-centric and Information-rich health care. Framework for strategic action. [Website]. Available from: <www.nhs.gov/healthit/documents/ hitframework.pdf> [Accessed: 3 June, 2009].

[36] Toohey, J. B., Herrick, J. D., and Trautman, R. T. (1982): Adaptation of a workload measurement system. *American Journal of Hospital Pharmacy* 39: 999-1004.

[37] Vemuri S. (1984): Simulated analysis of patient waiting time in an outpatient pharmacy. *American Journal of Hospital Pharmacy* 41: 1127-30.

[38] Wild, R. (1999). Essentials of Production and Operations Management. 5th ed. Great Britain" Wiltshire: Redwood Books.

[39] Winslade, N. E., Strand, L. M., Pugsley, J. A. and Perrier, D. G. (1996). Practice Functions Necessary for the Delivery of Pharmaceutical Care. *Pharmacotherapy,* 16 (5): 889 – 898.

Chapter 5

Improving Mandatory Environmental Data Reporting for Comparable and Reliable Environmental Performance Indicators

Mahelet G. Fikru

Additional information is available at the end of the chapter

1. Introduction

Since the 1980s mandatory national environmental data reporting has become one of the predominant environmental regulations in developed countries. Mandatory disclosure of environmental data usually involves creating a publicly accessible database or register whereby regulated entities periodically track, document and report environmental data such as emission of a list of priority pollutants and their transport to other sites [1]. All OECD Member States require industrial facilities to report the amount and type of pollutants released to air, water or land and wastes transferred off-site [2].

Mandatory environmental data reporting provides invaluable information that can be used to design policies, monitor and compare the environmental performance of companies as well as industries, improve cleaner production programs, reduce releases of certain chemicals, increase stakeholder participation and public awareness, increase accountability of organizations and address global environmental issues [1,3-7].

In addition, mandatory reporting reduces market failure resulting from insufficient disclosure of information on a good that produces negative environmental externalities [8, 9]. The study in [10] argues that mandatory reporting increases welfare compared to other instruments, such as optimal taxation, if implemented in a cost-effective way. Furthermore, several studies show that disclosure of environmental information affects the capital market and subsequently firm behavior whereby information about the poor environmental performance of publicly traded companies reduces their stock price return [11-14]. This is because poor environmental performance indicates high environmental related costs and/or liability which reduce the value of the company. The stock market reaction will subsequently induce a change in firm behavior whereby companies seek to reduce emission and improve environmental performance [15, 16].

For any given mandatory environmental register to harness the above benefits and more, the register has to be designed and implemented with cost-effectiveness and sustainability considerations [7]. In addition, it has to be constantly and extensively evaluated to check whether objectives are realized, policy needs are addressed, compliance is enforceable and any inconsistencies are minimized. Designs should also be improved to accommodate an expansion of geographical and sectorial coverage, an increase in the list of priority pollutants and better measurement techniques [17, 18]. Data should be standardized so as to allow comparison across companies, industries and over time. Once the above conditions are met and a reliable register is setup, reported data should easily reflect risk to human health and the environment so as to be easily interpreted by different stakeholders such as scholars, policymakers, managers and the public [19].

The purpose of this study is to explore and evaluate the most recent of all mandatory environmental data registers, namely the European Pollutant Release and Transfer Register or E-PRTR. The E-PRTR was adopted in 2006 by the European Parliament and the Council of the European Union with the purpose of making facility-level environmental information publicly available [20]. It obliges industrial facilities operating in 32 European countries (27 EU Member States, Iceland, Liechtenstein, Norway, Serbia and Switzerland) to report data on the release of harmful pollutants to air, water and land as well as their transport to other local and foreign sites. So far, data for three years (2007 to 2009) is publicly available for over 29,000 industrial facilities releasing or transferring 91 priority pollutants. The E-PRTR database is maintained by the European Commission and the European Environmental Agency and freely available at http://prtr.ec.europa.eu/ [21].

The E-PRTR has not yet been extensively evaluated to identify weaknesses, potential strengths and improvement areas. The only reviews available are the three informal reviews prepared by the European Environmental Agency with 'the objective of assisting countries to improve data quality by providing feedback on potential quality issues and inconsistencies with other reports'. In effect, the reviews are summaries and descriptive statistics of the 2007, 2008 and 2009 E-PRTR datasets [22-24]. The reviews also include a comparison of environmental data reported in the E-PRTR with other external environmental databases; and document errors and inconsistencies.

However, the reviews fail to address important issues like: How can reporting requirements be improved to be more inclusive? Are there any inconsistencies in reporting requirements and reported data? How can the E-PRTR be used for economic, environmental and policy analyses and what are things to consider when using the E-PRTR in such studies? How can the E-PRTR be used to study and compare the environmental impact of industrial sites, even if they operate in different countries? How can the E-PRTR be used to develop measures that reflect waste recycling and treatment efforts?

The contribution of this chapter is to address the above research questions and examine factors that government officials, businesses, academicians and citizens should consider when reviewing or using the E-PRTR. The primary purpose of the chapter is to identify limitations and strengths of the E-PRTR and recommend areas of improvement. Identification of potential strengths and weaknesses will assist regulators in future policy actions as well as policy design;

it will also encourage scholars to apply the freely available E-PRTR dataset in economic, scientific and environmental researches.

Secondly, the chapter introduces a new methodology to aggregate and normalize facility-level environmental data obtained from the E-PRTR. The normalized values will then be used to construct an *environmental performance indicator* which can easily be used to compare and rank industrial facilities across time, industry and country. Facility-level rankings within an industry can be used for policy implementation such as emission trading and allocation of quotas; industry-level rankings can be used to study the relative pollution-intensity of sectors; country-level comparisons can be used to design international environmental agreements. The environmental performance indicator introduced in this chapter captures a facility's environmental impact by reflecting abatement efforts through waste recycling and pollutant treatment techniques.

In section 2, an overview of the E-PRTR dataset is presented. In section 3 aggregation and normalization techniques are discussed and a novel measure of the environmental performance of facilities is introduced. Finally, section 4 concludes by forwarding some recommendations on what to consider when using the E-PRTR for policy and economic analyses; and how to improve reporting requirements and future data collection.

2. E-PRTR overview

The E-PRTR was adopted in 2006 by the European Parliament and the Council of the European Union with the purpose of making facility-level environmental information publicly available [20]. It replaces and improves the previous European Pollutant Emission Register which provides environmental data for the years 2001 and 2005. The E-PRTR is publicly available and can be accessed at http://prtr.ec.europa.eu/ [21]. Industrial facilities operating in 32 European countries (27 EU Member States, Iceland, Liechtenstein, Norway, Serbia and Switzerland) have reported annual data from 2007 to 2009. This chapter is based on the recently available 2009 data. In 2009, over 29,000 industrial facilities operating in 32 countries have reported to the Register. About 80% of these reporting facilities operate in 8 countries namely, UK, Germany, Spain, France, Italy, Poland, Belgium, Czech Republic and the Netherlands.

Mandatory national environmental data reporting is usually accompanied by a document which describes factors to consider when using such datasets. For instance, the US Toxic Release Inventory (TRI), which is one of the oldest and most successful mandatory disclosure requirement, is accompanied by official documents and reports which discuss factors to consider when using the TRI, limitations of available data, tools for analyzing and interpreting data, etc. [25]. However, the relatively new E-PRTR has not yet produced a document which provides caution about how to interpret and analyze data. The *Guidance Document to the Implementation of the E-PRTR* [20] is only an instruction document describing reporting requirements, characteristic of regulated facilities, what and how to report data. There are no suggestions for researchers and policymakers on how to use and interpret the data.

In this section an overview of the E-PRTR is presented along with a discussion of major limitations, potential strengths and factors to consider when using the E-PRTR data. In section 2.1 a summary of the characteristics of facilities required to report is presented along with some notes of caution while using the E-PRTR for policy design and national studies. In section 2.2 the type of required information is presented and evaluated. Section 2.3 discusses some inconsistencies between E-PRTR reporting requirements and actual reported data.

2.1. Characteristics of regulated facilities

In 2009, a total of 29,196 facilities reported to the E-PRTR out of which 938 facilities have more than one location while the rest 28,259 have a unique location. Industrial facilities engaged in 9 activities as their primary sectors are required to report to the E-PRTR. These 9 sectors, referred to as "Annex I activities" in the *Guidance Document*, are presented in Table 1 along with number of reporting facilities in each sector in 2009 [20].

Facilities engaged in the 9 activities as their primary sector are obliged to report to the E-PRTR only if their production (or processing) capacity exceeds a given annual threshold specific to each sector [20]. This indicates that reporting facilities are more or less larger facilities where small and medium facilities are not required to report. As a result, the E-PRTR would only be suitable for studies that focus on environmental damages from large point sources.

Sector	Number of reporting facilities
Intensive livestock production & aquaculture	6,104
Waste & wastewater management	7,653
Production & processing of metals	4,296
Chemical industry	2,821
Mineral industry	2,196
Energy sector	2,011
Animal & vegetable from food & beverage sector	1,990
Other activities	1,276
Paper & pulp; wood production & processing	849
Total	29,196

Table 1. Annex I activities and reporting facilities (2009)

In addition, reporting facilities represent a very small fraction of total active enterprises in most sectors and countries. For instance, consider the manufacturing sector which consists of all production activities in Table 1 except waste and wastewater management and intensive livestock and aquaculture. As can be seen from Table 2, at the national level, reporting facilities account for a very small percentage of total manufacturing enterprises active in 2009.[1] This narrow E-PRTR coverage may be because most sectors are dominated by small and medium scale facilities which are not required to report.

---

[1] Data for total population of manufacturing enterprises active in 2009 is obtained from Eurostat statistical database [26].

Country	Number of facilities in the manufacturing sector		
	Total population	Reporting facilities	E-PRTR coverage (%)
Austria	28,223	154	0.55
Belgium	38,462	525	1.36
Bulgaria	33,143	96	0.29
Cyprus	6,494	12	0.18
Czech Rep.	153,019	506	0.33
Denmark	18,336	223	1.22
Estonia	7332	45	0.61
Finland	28,401	314	1.11
France	234,398	2,164	0.92
Germany	263,464	2,703	1.03
Hungary	51,803	278	0.54
Ireland	12,776	182	1.42
Italy	444,564	1,630	0.37
Latvia	7,636	15	0.20
Lithuania	13,679	34	0.25
Luxembourg	851	23	2.70
Netherlands	53,717	510	0.95
Norway	18,704	229	1.22
Poland	233,308	974	0.42
Portugal	74,234	324	0.44
Romania	54,652	181	0.33
Slovakia	60,330	146	0.24
Slovenia	17,672	123	0.70
Spain	227,607	1,694	0.74
Sweden	55,767	374	0.67
UK	149,840	1,650	1.10

Table 2. E-PRTR 2009 coverage

As a result, reporting facilities are not a good representative of the larger population and the E-PRTR may not be suitable for broader policy analysis and national studies. This is because non-reporting facilities altogether could possibly have a higher aggregate environmental impact compared to regulated facilities.

2.2. Data facilities are required to report

Regulated facilities are required to report the name of their facility, parent company if any, location, full address, main economic activity, the release and transfer of 91 priority pollutants and name and address of competent authority of the country of operation. The 91 priority pollutants are classified into 7 groups: chlorinated organic substances, greenhouse gases, heavy metals, inorganic substances, other organic substances, pesticides and other gases. The release and transfer of each of the 91 pollutants should be reported if release and transfer exceeds a given annual threshold specific to each pollutant.

Specifically, regulated facilities are required to annually report the amount of: (1) Pollutants released to air, water and land in kilogram given each pollutant released exceeds a given annual reporting threshold. Accidental releases should be reported separately whenever available. (2) Off-site transfer of solid waste (hazardous and non-hazardous) for the purpose of disposal in tons given that the transfer exceeds 2 tons of hazardous waste and 2000 tons of non-hazardous waste. (3) Off-site transfer of solid waste (hazardous and non-hazardous) for the purpose of recovery in tons given that the transfer exceeds 2 tons of hazardous waste and 2000 tons of non-hazardous waste. (4) Off-site transfer of pollutants in wastewater (through pipes) for the purpose of treatment in kilogram given that the transfer exceeds a given annual reporting thresholds for each pollutant. Whenever hazardous and non-hazardous wastes are transferred off-site to another country for recovery or disposal, the name and address of the receiving facility along with purpose of transfer should be reported.

Table 3 reports a distribution of number of facilities with the type of required data. Close to 60% of regulated facilities have transferred hazardous and/or non-hazardous wastes off-site, possibly to a specialized waste handler for the purpose of recovery or disposal. On the other hand, only 6% of regulated facilities have transferred pollutants in wastewater through pipes to external waste handlers for the purpose of treatment.

Reported data	Reporting facilities
Pollutant released to air, water & land	14,170
Off-site transfer of hazardous & non-hazardous waste for recovery	17,363
Off-site transfer of hazardous & non-hazardous waste for disposal	17,190
Off-site transfer of pollutants in wastewater for treatment	1,769

Table 3. Data facilities are required to report

Regulated facilities are not required to identify the different types of hazardous (and non-hazardous) wastes they transfer to external waste handlers. Rather, the amount of all types of hazardous wastes is reported as an aggregated value. As a result, there are no mechanisms to differentiate facilities which generate 'acutely' hazardous wastes from facilities which generate 'slightly' hazardous wastes.

In addition to reporting the amount of different pollutants and wastes released and transferred from individual facilities, the E-PRTR requires facilities to report the techniques used to determine reported values. Accordingly, facilities clearly indicate values directly measured and values calculated based on production or input data. When amounts are calculated or measured the technical method used to measure and calculate should be reported [20]. If direct measurements and calculations are not possible then estimations based on professional assumption are allowed. As Table 4 shows only a few values have been estimated and close to 90% of data are based on either calculations or actual measurements. Measured and calculated values are more accurate than estimated values and hence the E-PRTR dataset can be considered as a reliable source.

Reported data	Calculated	Estimated	Measured
Pollutant releases	46%	10%	44%
Pollutants transferred off-site in wastewater	15%	12%	73%
Off-site transfer of hazardous & non-hazardous waste	54%	0%	46%

Table 4. Reporting techniques

Facilities should disclose and report all required information to the competent authority of their country, which is responsible for passing the information to the European Commission. Whenever facilities have justifiable reasons they can request confidentiality not to disclose some information. Reasons for confidentiality should be based on Article 4 of Directive 2003/4/EC of the European Parliament and the Council of 28 January 2003 on public access to environmental data. The type of information withheld with the reason of withholding should be reported to the public. Some of the justifications for declaring confidentiality are: disclosure compromises public security, international relations, the ability of a person to receive fair trial, legitimate economic interest, intellectual property right, protection of location of rare species, etc. [27]. A total of 125 facilities have declared confidentiality and these are located in Belgium (106 facilities), Bulgaria (3 facilities), Denmark (1 facility), Germany (3 facilities), Luxembourg (3 facilities), Romania (2 facilities), Sweden (4 facilities) and UK (2 facilities). These confidential firms are engaged in animal and vegetable production as food and beverage (4 facilities), chemical industry (39 facilities), energy (6 facilities), mineral (4 facilities), paper and wood (1 facility), production and processing of metals (21 facilities), waste management (46 facilities) and others (4 facilities). Table 5 presents the number of facilities declaring confidentiality based on Article 4(2) of Directive 2003/4/EC.

Despite presenting fairly accurate environmental data, the absence of other important and complementary variables limits the usefulness of the E-PRTR dataset in comparative analysis. This chapter identifies at least four types of potentially useful information that are currently missing from the E-PRTR but can easily be incorporated by updating reporting requirements.

Confidential information	Number of facilities
Is off-site transfer of waste destined for recovery or disposal?	
Off-site transfer of hazardous waste (inside country)	1
Off-site transfer of hazardous waste (outside country)	3
Off-site transfer of non-hazardous waste	6
Is reported value based on measurement, calculation or estimation?	
Off-site transfer of hazardous waste (inside country)	3
Off-site transfer of hazardous waste (outside country)	1
Off-site transfer of non-hazardous waste	5
Information on host for off-site transfers	
Waste handler party name & address	124
Quantity of hazardous & non-hazardous waste transferred offsite	13
Total	125

Table 5. Confidentiality

First of all, the E-PRTR questionnaire includes sections where facilities can report optional but not required information on facility characteristics. For instance, data on production volume, number of installations, number of operating hours, number of employee data are only optional and not required. As a result only 2.6% of facilities have reported production quantities and only 9% have reported number of employees' data. Providing space for reporting the above information but making it optional is a drawback of the E-PRTR. This is because production and input data provide information on firm size which could easily be used to normalize emission and transfer data and construct easily comparable environmental performance indicators which control for size.

Secondly, essential input-use variables are not required by the E-PRTR at all. For instance, the amount of harmful chemical inputs, raw materials consumed, energy consumed, total waste generated and number of permits held, if any, are not required by the E-PRTR. This is another limitation of the E-PRTR since input variables could assistant the construction of a broad range of alternative environmental performance indicators. For instance, one could construct indicators based on emission per production or per number of employees to measure environmental impacts or else construct energy use per production to measure efficiency of resource use.

Thirdly, there are no provisions that allow the identification of firms that use cleaner production technologies, if any. Available data on the transfer of pollutants in wastewater through pipes for treatment and the transfer of hazardous and non-hazardous wastes for recovery and disposal only indicate abatement efforts through end-of-the-pipe-type techniques. Even though cleaner technologies solve environmental problems better by preventing rather than just treating pollution, the E-PRTR provides no mechanisms to identify and reward regulated facilities which rely on these technologies [28, 29]. Future data collection can easily be improved by including questions to identify whether facilities are engaged in any source reduction activities or activities that use less natural resources.

Fourthly, the E-PRTR has no provisions for reporting abatement that takes place on-site by using end-of-the-pipe techniques. The focus of the E-PRTR is on reporting the transfer of pollutants to specialized off-site waste management facilities. For instance, facilities are not required to report on-site recycling, energy recovery and treatment. Figure 1 illustrates how the absence of on-site waste management and other variables from the E-PRTR can create biases and reduce its usefulness. The figure is adopted from the *Guidance Document for the Implementation of the E-PRTR* and modified to illustrate the limitation of E-PRTR reporting requirements [20].

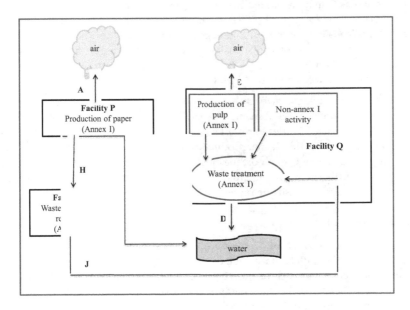

Figure 1. E-PRTR requirements and limitations

The three facilities illustrated in Figure 1 (Facility P, Q and R) are engaged in Annex I activities as their primary sector. All three facilities exceed the production threshold for their respective activity and hence are regulated by the E-PRTR. Facility Q owns a wastewater treatment plant which is also a part of Annex I activities. Facilities P and R transfer waste to facility Q for treatment and this should be reported as off-site transfer of pollutants in wastewater for treatment as amounts C and J respectively. In addition, facility P should report amount A and B as pollutant released to air and water respectively; amount H as off-site transfer of wastes. Facility Q should report amount E and D as pollutant released to air and water respectively.

At least two drawbacks of the E-PRTR reporting requirements can be illustrated using Figure 1. First of all, amounts F and G, which measure on-site waste management, remain unreported. Amount G is not reported since it originates from a non-Annex I activity. However, even though amount F originates from an annex I activity, it is not reported. It may appear as if facility Q is not abating wastes at all since it has no transfers for treatment, disposal or recovery. As a result, the E-PRTR cannot be used to identify and reward facilities which invest in their own waste management plants.

Secondly, there may be double counting of wastes in amounts H and J (assuming all amounts A to H exceed reporting thresholds). It may be the case that part of the waste that originated as H from facility P is not processed by facility R and hence ends up as J to facility Q. Take another example, the amount reported as C originates and is reported by facility P. Part of the amount indicated as C may be treated by facility Q while the rest ends up as E or D in the treatment process. However, the E-PRTR records C, E and D as if they are independent amounts and this leads to double counting. In general, the E-PRTR does not follow the fate of a given pollutant from generation to disposal or recovery and there are no mechanisms to trace the source of pollutants and wastes. As a result, the E-PRTR does not enable us to understand the true relationship between regulated facilities.

Some of the above discussed drawbacks of the E-PRTR can easily be improved by drawing lessons from other more successful mandatory environmental registers such as the US TRI. For instance, by requiring facilities to report source reducing and other activities the US TRI can better identify facilities with cleaner production. In addition, the US TRI identifies amounts F and G by requiring facility Q to report the amount of wastes/pollutants recycled, recovered or treated on-site. Furthermore, the US TRI has reporting requirements on on-site disposal and storage of wastes which can easily be replicated by the E-PRTR.

2.3. Reporting thresholds

Facilities which exceed given production (or processing) capacity thresholds in each sector are required to report to the E-PRTR only if the release and transfer of pollutants exceeds a given annual threshold specific to each of the 91 priority pollutant [20]. Reporting threshold for pollutants is given for each of the three medium of releases as air, water and land. The annual reporting threshold is given as 2 tons for hazardous wastes and 2000 tons for non-hazardous wastes. Regarding reporting thresholds this chapter identifies two inconsistencies between reporting requirements and actual reported data; and one source of lack of transparency in the construction of reporting thresholds.

Pollutant release	Missing threshold for	Number of reporting facilities
Phenols	air	6
Fluorides	air	18
1-1-1- Tricholorethhane	water	6
Sulfur oxides	water	2
Total nitrogen	air	3
Total organic carbon	air	57
Total organic carbon	land	14
Other non-priority pollutants	-	10

Table 6. Missing thresholds

First of all, facilities are required to report to the Register only if given reporting thresholds are exceeded. However, some facilities have reported releases and transfers below the reporting threshold. For instance, a total of 311 facilities reported all values below or equal to reporting thresholds; these facilities were not meant to be regulated by the E-PRTR. This inconsistency between reporting requirements and actual data raises concerns on the effectiveness and uniformity of the implementation of the E-PRTR.

The second inconsistency regarding reporting thresholds is that some facilities reported releases of certain pollutants even though they have no reporting threshold and hence do not trigger reporting. Table 6 summarizes number of facilities which reported data despite missing threshold requirements. Whenever firms report unrequired data it is not clear what thresholds they used, if they used one. This creates inconsistency and reduces the comparability of data obtained from the E-PRTR.

The possibility of missing thresholds indicates that the E-PRTR has a potential to expand the list of priority pollutants as well as their medium of releases. For instance, emission of toluene, xylenes, fluorides, atrazine, phenols and benzo perylene to air does not trigger a reporting requirement. However, these pollutants have known human toxicity when released to air and hence should be documented and reported [30].

Another concern regarding reporting thresholds is that the basis on which the thresholds were set is neither transparent nor clearly explained. Reporting thresholds should ideally reflect the impact of pollutants on human health and the environment. In other words, highly toxic pollutants should have a lower threshold compared to less toxic pollutants. Without an expert toxicology analysis, the thresholds for most of the 91 priority pollutants seem to be based on effects and impact. It appears like 'more harmful' substances have lower thresholds than 'less harmful' substances. Nevertheless the European Commission and European Environmental Agency have not yet given a detailed explanation on how the thresholds are set [20].[2] The only

---

2 It is anticipated that the European Commission will carry out an evaluation of pollutant thresholds and publish a review in 2013.

indication given on the E-PRTR website is that "thresholds have been set with the intention of covering for each specific pollutant about 90% of the total mass emissions from facilities regulated under E-PRTR" [21]. Thresholds for 50 pollutants were decided in 1998-99 but never updated since; at the time not much was known about total emissions in Europe. For the rest 41 pollutants, 'pragmatic solutions' were adopted to incorporate into the E-PRTR Regulation [31]. However, the E-PRTR should not just be about total releases into the environment from large point sources but also about harm and risk per se.

Some pollutants are long-lived and persist in the environment for a longer period of time (consider benzene which can persist in the environment for a week) while others are short lived like Hydrogen Chloride which persists for about 7 hours [30]. In addition the impact of pollutants depends on the geographical and atmospheric condition, height of release, transport of pollutant as well as population density. Thus, thresholds should consider impact factors and risk rather than just considering gross emissions. For instance, copper and chromium have equal reporting thresholds of 100kg released to air. But 1 gram of chromium requires a higher volume of air of about 1 million m^3 to loss its human toxicity (through inhalation) as compared to copper which requires only 570 m^3 of air. Not only human toxicity but other impacts like ecotoxicity should be considered when setting the reporting thresholds.

Despite some of its limitations, the E-PRTR provides a framework for collecting rich facility-level environmental data for several countries. Hence, it has the potential to address the problem of lack of facility-level data on industrial wastes since most other currently available data are dominated by municipal wastes [32]. In addition, the E-PRTR reports environmental data from several countries using the same reporting rules; this provides a potential basis for comparing international data and designing international environmental agreements. The E-PRTR can also be used to develop measures that reflect the waste recycling and treatment effort of facilities.

In section 3, a procedure for aggregating and normalizing E-PRTR data is presented. The purpose of normalization is to develop a comparable environmental performance indicator.

3. Environmental Performance Indicator (EPI)

The 91 priority pollutants released to air, water and land greatly differ in their effect on human health and the environment. They also differ in their toxicity, persistent in the environment, transport and fate. Because of this heterogeneity one cannot just add the amounts of the different pollutants reported by a facility. Rather one needs to normalize the raw data by using weights that reflect toxicity as well as impact on the environment.

As discussed in section 2.3, reporting thresholds should ideally reflect the impact of pollutants on human health and the environment where highly toxic pollutants are given a lower threshold compared to less toxic pollutants. Even though the European Commission provides no detailed explanation on how E-PRTR thresholds are set, it appears like 'more harmful' substances roughly have lower thresholds than 'less harmful' substances.

If E-PRTR thresholds appropriately reflect risk then one can use the thresholds to aggregate the several pollutants released and transferred by a single firm. This study introduces a normalization procedure whereby we calculate the percentage of a pollutant a facility has released or transferred over the given reporting thresholds. If reporting thresholds correctly reflect impact of pollutants on health and the environment, then this normalized value represents the environmental impact of a facility. A higher normalized value represents a higher impact, whereas a lower value represents a lower impact. For each firm we use the following normalization formula:

$$\sum_{p=1}^{q} \frac{X_p - T_p}{T_p} \qquad (1)$$

where X is the actual amount of pollutant p released or transferred off-site, T is the given reporting threshold for a pollutant p and q is the number of the different types of pollutants a firm has released or transferred in a year.

The above formula yields a unit-free number which can be interpreted as how much above the threshold a firm has emitted or transferred. The normalized value presented in equation (1) cannot by itself be used to compare firms. This is because large sized firms have a higher normalized value while smaller firms naturally have a lower value. However, the normalized data can be used to develop an *environmental performance indicator* which controls for the size of firms.

Environmental performance indicators are quantitatively measurable results of a facility's operation that interact with the environment [33]. Environmental performance has three dimensions: preventing waste before it occurs using cleaner production; recycling, treating or reducing waste using end-of-the-pipe techniques; and using resources and energy efficiently [34, 35].

Environmental performance indicators should be comparable across time and firms, target-oriented and understandable for users [33, 36]. Furthermore, indicators should be consistent with policymakers' and regional priority. In addition, normalization schemes should reflect environmental impacts/damages, should be easily replicable, transparent, easy to interpret, and available for all relevant pollutants [37]. Developing a performance indicator requires collecting accurate data, assessing information against objectives and criteria, selecting indicators, reporting and communicating results, reviewing and improving indicators. Several studies have developed numerous normalization techniques which reflect weighting, aggregating and comparing pollutants released in different mediums based on their health and environmental impacts [38-40]. Techniques significantly vary based on type of available data, type of firm-ownership (public versus private) and sector being considered [41, 42]. Commonly used indicators capture a company's effort in using end-of-the-pipe techniques and efficient input use. For instance, electricity consumed per production, quantity of waste generated per production and quantity of waste recycled per production are possible performance indicators [36, 43].

Based on data available from the E-PRTR, we introduce a firm-level environmental perform-ance indicator that controls for size of firms. The environmental performance indicator introduced in this chapter captures a firm's environmental impact and reflects abatement using end-of-the-pipe-type waste reduction efforts. Data and reporting requirements are not sufficient to explore efficiency of resource and energy use or abatement through cleaner technologies. Hence, the introduced indicator only reflects waste recycling and treatment efforts by using end-of-the-pipe techniques.

In this study, environmental performance of a firm is defined as the amount of waste treated and recovered as a percentage of total wastes/pollutants where total wastes/pollutants is the sum of waste disposed, recovered, treated and pollutants released to air, water and land. The following formula is used to calculate the environmental performance of any firm (say firm i) where all variables are normalized values

$$EPI_i = \frac{W_{t,i} + W_{r,i}}{e_i + W_{t,i} + W_{r,i} + W_{d,i}}$$

(2)

where EPI_i is the environmental performance indicator for firm i, e_i is normalized value for pollutants released to air, water and land, $W_{r,i}$ is normalized value for waste recovered, $W_{t,i}$ is normalized value for waste treated and $W_{d,i}$ is normalized value for waste disposed.

EPI_i yields a value between 0 and 1 where firms with EPI close or equal to 1 have 'good' performance as they have succeeded to abate 100% of pollution and wastes using end-of-the-pipe techniques. On the other hand, firms with EPI close to 0 have 'bad' performance as they have not used end-of-the-pipe techniques to treat pollutants and recover wastes. EPI values can be converted to percentages where $EPI_i = x\%$ can be interpreted as 'firm i has abated $x\%$ of pollutants and wastes using end-of-the-pipe techniques'.

We use equation (2) to calculate EPI for facilities regulated by the E-PRTR. For a total of 311 facilities the calculated EPI is undefined because all of the 4 normalized values used to construct equation (2) are zero. If a given facility has all zeros for all normalized values, then that facility should not have reported to the E-PRTR as it does not exceed reporting thresholds. To maintain comparability we take out the 311 facilities with undefined EPI values and report a ranking of environmental performance for the remaining facilities. See Table 7.

As Table 7 indicates, close to 59% of facilities have 'bad' environmental performance with EPI less or equal to 25%. This indicates that large point sources are not investing much on treatment of pollutants and recycling of wastes off-site. Nevertheless whether these sources are relying on cleaner technologies or not is not clear from available data. It is also not clear whether they rely on on-site waste management and recycling rather than using off-site options.

EPI %	Percentage of facilities
0%	43.38%
0 % - 25%	15.23%
25% - 50%	7.70%
50% - 75%	6.99%
75% - 99%	11.70%
100%	15.00%

Table 7. Environmental performance of firms

The environmental performance indicator presented in equation (2) can be extended at the country level by using the following formula:

$$EPI_k = \frac{\sum_{i=1}^{n} W_{t,i} + \sum_{i=1}^{n} W_{r,i}}{\sum_{i=1}^{n} e_i + \sum_{i=1}^{n} W_{t,i} + \sum_{i=1}^{n} W_{r,i} + \sum_{i=1}^{n} W_{d,i}} \qquad (3)$$

where EPI_k is the environmental performance indicator of country k and n is the number of regulated facilities in each country. Equation (3) makes use of the sum of normalized values at the country level.

At the country level, EPI_k measures the overall performance of facilities regulated by the E-PRTR where good performance is in terms of abating a larger percentage of wastes and pollutants using off-site end-of-the-pipe techniques. The EPI_k measure controls for number of regulated firms in each country as well as firm size.

Table 8 presents a ranking of countries based on equation (3). Countries such as Latvia, Lithuania, Malta, Iceland and Bulgaria have the highest environmental performance indicator. High environmental performance in these countries may be because they have high investment on off-site end-of-the-pipe technologies of recycling and treatment. The largest Member States like Germany and France also have relatively good performance ranking at 7th and 15th places with 65% and 47% EPI values respectively. On the other hand, countries such as UK and Switzerland are on the lower end with EPI values of 21% and 20% respectively. This may be because these countries rely more on cleaner production and other on-site waste management techniques rather than using off-site end-of-the-pipe-type abatement options.

One can also construct a sector level environmental performance indicator to identify sectors which use off-site end-of-the-pipe techniques for treatment and recovery of wastes and pollutants. Table 9 presents a sector level ranking based on the environmental performance indicator. 'Waste and wastewater management' as well as 'paper, pulp and wood production and processing' industries have the highest EPI value of above 50%. On the other hand the 'mineral industry' and 'intensive livestock production and aquaculture' have the lowest EPI values.

	Reporting facilities		
Country	Number	Percentage	EPI_k(%)
Latvia	32	0.11	91.63
Lithuania	97	0.34	81.14
Malta	15	0.05	74.14
Iceland	17	0.06	72.47
Bulgaria	182	0.63	71.28
Luxembourg	29	0.10	69.70
Germany	4,692	16.24	65.25
Poland	1,287	4.46	58.69
Austria	239	0.83	57.32
Romania	484	1.68	53.85
Ireland	332	1.15	52.68
Sweden	549	1.90	51.26
Denmark	425	1.47	48.76
Belgium	914	3.16	48.19
France	3,563	12.34	47.68
Cyprus	66	0.23	44.41
Czech Rep.	800	2.77	39.65
Spain	3643	12.61	38.13
Portugal	574	1.99	36.98
Netherlands	790	2.73	35.34
Slovenia	183	0.63	30.34
Hungary	730	2.53	28.48
Italy	2,582	8.94	27.81
Norway	724	2.51	21.11
UK	4,713	16.32	21.03
Switzerland	222	0.77	20.20
Finland	480	1.66	15.03
Estonia	101	0.35	14.17
Greece	124	0.43	8.78
Slovakia	256	0.89	8.76
Liechtenstein	1	0.00	0.00
Serbia	39	0.14	0.00

Table 8. Environmental performance of countries

The *EPI* measure introduced in this paper is not without limitations. Although most of the weaknesses of the *EPI* measure are inherent in the limitations of existing E-PRTR reporting requirements, there are some limitations we wish to acknowledge in the application of the *EPI* in future studies. First of all, even if thresholds correctly and fully reflected risk, it may not fully capture the potential risk of pollutants released to air, water and land since *EPI* is constructed based on the presence of a given amount of a pollutant in the environment (air, water, land). We have no information on exposure to humans, how long pollutants stay in the environment and how many people live around a given facility. Secondly, even though different thresholds are given for different medium of releases (i.e. air, water and land), the medium of releases are treated equally and given identical weights when constructing *EPI*. Lastly, *EPI* does not reflect efficiency of resource/energy use and cleaner production efforts since available data is not sufficient to explore these.

Sectors	Reporting facilities (%)	EPI %
Waste & wastewater management	26.21	54.40
Paper, pulp & wood production processing & processing	2.91	51.49
Other activities	4.37	45.86
Production & processing of metals	14.71	42.58
Energy sector	6.89	32.45
Animal & vegetable products from food &beverage	6.82	29.34
Chemical industry	9.66	22.11
Mineral industry	7.52	18.11
Intensive livestock production & aquaculture	20.91	7.26

Table 9. Environmental performance of sectors

4. Conclusion and recommendations

The recently introduced Europe-wide mandatory environmental data reporting regulation, known as E-PRTR, has not yet been extensively evaluated to assess weaknesses and inconsistencies that prevent it from being widely used in economic and policy analysis. There are yet no documents that provide caution on how to use, interpret and analyze data obtained from the E-PRTR.

This chapter explores this relatively new database and underlines some factors to consider when using the E-PRTR in academic research or policy design. This study recognizes that the E-PRTR would only be suitable for researches that focus on environmental damages from large point sources. This is because reporting requirements exclude small and medium sized facilities with production capacity below a given threshold. In addition, reporting facilities represent a very small fraction of total active enterprises. Hence, reporting facilities may not be a good representative of the larger population; this limits the use of E-PRTR for broader policy analysis and national studies.

This chapter identifies some of the major limitations of the E-PRTR. Important and complementary variables are missing from the E-PRTR regulation which limits the usefulness of the E-PRTR dataset in comparative studies. Data on production volume, number of installations, number of operating hours and number of employee data are only optional and hence a very small number of facilities have responded. These data are essential in that they provide information on firm size which could have easily been used to construct easily comparable environmental performance indicators. Other input use variables such as the amount of harmful chemical inputs, raw materials consumed, energy consumed, total waste generated and number of permits held, if any, are not required by the E-PRTR. Such variables could assistant the construction of a broad range of alternative environmental performance indicators. Furthermore, reporting requirements do not identify facilities which use cleaner production technologies and facilities which use on-site recycling, energy recovery and treatment. As a result, the E-PRTR cannot be used for policies that reward facilities which invest in their own waste management plants. In addition, reporting requirements do not prevent double counting of wastes since there are no mechanisms to trace pollutants and wastes from cradle to grave. Another limitation of the E-PRTR is that it provides no mechanisms to differentiate facilities which generate 'acutely' hazardous wastes from facilities which generate 'slightly' hazardous wastes.

The chapter also identifies some inconsistencies between reporting requirements and actual reported data. A few facilities which were not meant to be regulated by the E-PRTR have reported data whereas other regulated facilities have reported pollutants which do not trigger reporting. Such inconsistencies between reporting requirements and actual reports raise concerns on the effectiveness and uniformity of the implementation of the E-PRTR as well as comparability of data. Besides these inconsistencies, there are no detailed explanations on how pollutant reporting thresholds have been constructed. As it is, it seems like these thresholds are just rough estimations.

In addition to exploring the E-PRTR the chapter also introduces a new methodology to aggregate and normalize facility-level environmental data. Normalized values are used to construct an environmental performance indicator which captures a facility's environmental impact by reflecting abatement efforts through waste recycling and pollutant treatment techniques. The indicator can easily be used to compare the environmental impact of industrial facilities across time, industry and country.

Finally, based on the evaluation of the E-PRTR summarized above, we forward the following recommendations to improve future data collection and existing reporting requirements.

i.   Reporting requirements can be improved to be more inclusive and representative by lowering production capacity thresholds and increasing the list of priority pollutants to include chemicals with known harm to health and the environment.

ii.   Future data collection can be improved by making mandatory the disclosure of input and output variables such as energy use, production volume, raw material use, total waste generated, etc.

iii.   Facilities which rely on cleaner production should be identified. Similar to the US TRI the E-PRTR can oblige facilities to report source reducing activities.

iv.   Similar to the US TRI, facilities can also be required to report the amount of wastes/ pollutants recycled, recovered or treated on-site as well as the amount of wastes/ pollutants disposed or stored on-site.

v.   Inconsistencies can be reduced by making reporting requirements more standardized and ensuring uniformity of their implementation. For instance, there should be mechanisms to prevent non-regulated firms from reporting and to ensure that regulated firms are actually reporting according to given guidelines.

vi.   Further study on the toxicity of priority pollutants is required to understand, categorize and rank the effect of each priority pollutant on human health and the environment. This would help the construction of transparent reporting thresholds which are based on risk and impact.

vii.   Waste categories such as hazardous and non-hazardous wastes can further be decomposed to differentiate the content of waste. For example, the US Environmental Protection Agency categorizes hazardous wastes into manufacturing wastes, wastes from specific industries, wastes from commercial chemical products, mixed wastes containing hazardous and radioactive components, pesticides, etc.

Author details

Mahelet G. Fikru

Address all correspondence to: fikruma@mst.edu

Department of Economics, Missouri University of Science and Technology, USA

References

[1] Case D. The Law and Economics of Environmental Information as Regulation. Environmental Law Institute 2001; 31 ELR 10773. http://www.vanderbilt.edu/vcems/papers/ELRVersion2.pdf (accessed 10 May 2012).

[2] Pollutant Release and Transfer Register. www.prtr.net/en/ (accessed 2 April 2012).

[3] OECD. Pollutant Release and Transfer Registers (PRTRs). A Tool for Environmental Management and Sustainable Development. PRTR Workshop for the America's, July 29-31, 1997, San Juan del Rio, Mexico. http://www.oecd.org/dataoecd/ 36/32/2348006.pdf (accessed 13 April 2012).

[4] Larrinaga C., Carrasco F., Correa C., Llena F., Moneva J. Accountability and Accounting Regulation: The Case of the Spanish Environmental Disclosure Standard. The European Accounting Review 2002; 11(4) 723-740.

[5] Kolominskas C., Sullivan R. Improving Cleaner Production through Pollutant Release and Transfer Register Reporting Processes. Journal of Cleaner Production 2003; 12(7) 713-724.

[6] National Environment Protection Council. National Environment Protection (National Pollutant Inventory) Measure. Australia 1998. http://www.comlaw.gov.au/Details/ F2007B01123 (accessed 13 April 2012).

[7] Fung A., Graham M., Weil D. The Political Economy of Transparency- What Makes Disclosure Policies Sustainable? Ash Institute for Democratic Governance and Innovation, John F. Kennedy School for Government, Harvard University; 2004. http:// www.transparencypolicy.net/assets/whatnakesdisclosureeffective.pdf (accessed 4 April 2012).

[8] Kennedy P., Laplante B., Maxwell J. Pollution Policy: The Role for Publicly Provided Information. Journal of Environmental Economics and Management 1994; 26(1) 31-43.

[9] Kleindorfer P., Orts E. Information Regulation of Environmental Risk. Risk Analysis 1998; 18(2) 155-170.

[10] Petrakis E., Sartzetakis ES., Xepapadeas A. Environmental Information Provision as a Public Policy Instrument. Contributions to Economic Analysis & Policy 2005; 4 (1).

[11] Hamilton JT. Pollution as News: Media and Stock Market Reactions to the Toxics Release Inventory Data. Journal of Environmental Economics and Management 1995; 28(1) 98-113.

[12] Konar S., Cohen MA. Information as Regulation: The Effect of Community Right to Know Laws on Toxic Emission, Journal of Environmental Economics and Management 1997; 32 (1) 109-124.

[13] Konar S., Cohen MA. Does the Market Value Environmental Performance? The Review of Economics and Statistics 2001; 83(2) 281-289.

[14] Khanna M., Quimio WR., Bojilova D. Toxics Release Information: A Policy Tool for Environmental Protection. Journal of Environmental Economics and Management 1998; 36(3) 243-266.

[15] Al-Tuwaijri S., Christensen T., Hughes II KE. The Relations among Environmental Disclosure, Environmental Performance, and Economic Performance: A Simultaneous Equations Approach. Accounting, Organizations and Society 2004; 29(5-6) 447–471.

[16] Clarkson P., Li Y., Richardson G., Vasvari F. Does it Really Pay to be Green? Determinants and Consequences of Proactive Environmental Strategies. Journal of Accounting and Public Policy 2011; 30(2) 122-144.

[17] Kakkainen B. Information as Regulation: TRI and Performance Benchmarking, Precursor to a New Paradigm? Georgetown Law Journal 2001; 89(257) 1-89.

[18] Kerret D., Gray G. What Do We Learn from Emissions Reporting? Analytical Considerations and Comparison of Pollutant Release and Transfer Registers in the US, Canada, England and Australia. Risk Analysis 2007; 27 (91) 203-223.

[19] Tietenberg T. Disclosure Strategies for Pollution Control. Environmental and Resource Economics 1998; 11(3-4) 587-602.

[20] European Commission/EC/. Guidance Document for the Implementation of the European PRTR 2006. http://prtr.ec.europa.eu/docs/EN_E-PRTR_fin.pdf (accessed 13 April 2012).

[21] European Pollutant Release and Transfer Register. http://prtr.ec.europa.eu/ (accessed 12 July 2012).

[22] European Environmental Agency/EEA/. E-PRTR Review Report 2009 Covering the 2007 E-PRTR Dataset. ETC/ACC Technical Paper 2009/15. The Netherlands, 2010a.

[23] European Environmental Agency/EEA/. E-PRTR Review Report 2010 Covering the 2008 E-PRTR Dataset. ETC/ACC Technical Paper 2010/05. The Netherlands, 2010b.

[24] European Environmental Agency/EEA/. E-PRTR Informal Review Report 2011 Covering the 2009 E-PRTR Dataset. ETC/ACM Technical Paper 2011/06. The Netherlands, 2011.

[25] USA Environmental Protection Agency TRI. The Toxic Release Inventory (TRI) and Factors to Consider When Using TRI Data. http://www.epa.gov/tri/tridata/index.html (accessed 4 May 2012).

[26] European Commission Eurostat Statistical Database. http://epp.eurostat.ec.europa.eu/portal/page/portal/statistics/search_database (accessed 12 April 2012).

[27] Directive 2003/4/EC of the European Parliament and the Council of 28 January 2003. Official Journal of the European Union 2003; 41(-) 26-32. http://eur-lex.europa.eu/LexUriServ/LexUriServ.do?uri=OJ:L:2003:041:0026:0032:EN:PDF (accessed 7 February 2012).

[28] Fikru MG. Does the E-PRTR Enable us to Understand the Environmental Performance of Firms? Environmental Policy and Governance 2011; 21(3) 199-209.

[29] Johnstone N., Serravalle C., Scapecchi P., Labonne J. 2007. Public Environmental Policy and Corporate Behavior: Project Background, Overview of the Data and Summary Results. In: Johnstone N. (ed.) Environmental Policy and Corporate Behavior. UK: OECD p1–33.

[30] Hauschild M., Potting J. Spatial Differentiation in Life Cycle Impact Assessment- The EDIP2003 Methodology. Environmental News No. 80. Danish Ministry of the Environment 2005.

[31] Air and Climate Change – Mitigation Office. European Environmental Agency. Denmark .E-mail to Bob Boyce (Bob.Boyce@eea.europa.eu) 2011 October 18.

[32] OECD. Key Environmental Indicators 2008. http://www.oecd.org/dataoecd/20/40/37551205.pdf (accessed 5 May 2012).

[33] Perotto E., Canziani R., Marchesi R., Butelli P. Environmental Performance, Indicators and Measurement Uncertainty in EMS Context: A Case Study. Journal of Cleaner Production 2008; 16(4) 517-530.

[34] Cordano M., Frieze I. Pollution Reduction Preferences of US Environmental Managers: Applying Ajzen's Theory of Planned Behavior. Academy of Management Journal 2000; 43(3) 627-641.

[35] Klassen R., Whybark D. The Impact of Environmental Technologies on Manufacturing Performance. Academy of Management Journal 1999; 42(6) 599-615.

[36] Jasch C. Environmental Performance Evaluation and Indicators. Journal of Cleaner Production 2000; 8(1) 79-88.

[37] Larsen HF., Hauschild M. Evaluation of Eco-toxicity Effect Indicators for Use in LCIA. International Journal of LCA 2007; 12(1) 24-33.

[38] Hertwich E., Mateles S., Pease W., McKone T. Human Toxicity Potentials for Life-Cycle Assessment and Toxics Release Inventory Risk Screening. Environmental Toxicology and Chemistry 2001; 20(4) 928–939.

[39] Bare J., Norris G., Pennington D., McKone T. TRACI The Tool for the Reduction and Assessment of Chemical and other Environmental Impacts. Journal of Industrial Ecology 2003; 6(3-4) 49-78.

[40] Toffel M., Marshall J. Improving Environmental Performance Assessment. A Comparative Analysis of Weighting Methods used to Evaluate Chemical Release Inventories. Journal of Industrial Ecology 2004; 8 (1-2) 143-172.

[41] Ren X. Development of Environmental Performance Indicators for Textile Process and Product. Journal of Cleaner Production 2000; 8(6) 473-481.

[42] Lundberg K., Balfors B., Folkeson L. Framework for Environmental Performance Measure in a Swedish Public Sector Organization. Journal of Cleaner Production 2009; 17(11) 1017-1024.

[43] Rao P., Castillo O., Intal PS., Sajid A. Environmental Indicators for Small and Medium Enterprises in the Philippines: An Empirical Research. Journal of Cleaner Production 2006; 14(5) 505-515.

Technology Assessment in Software Development Projects Using a System Dynamics Approach: A Case of Application Frameworks

José Ignacio Muñoz Hernández,
José Ramón Otegui Olaso and
Alejandro Gutiérrez López

Additional information is available at the end of the chapter

1. Introduction

Technology is in constant change, new knowledge creates new technological opportunities; and there are always possibilities for improvement [1]. This constant innovation creates a plethora of technological alternatives from where an organization can choose and implement. In addition to this large array of alternatives, technologies are becoming more complex and evolving faster, making selection more difficult.

Technologies can create competitive advantages if they are selected correctly, this selection can be considered one of the most challenging decision making areas managers face [2]. The selections must match the organization previous technologies and systems, as well as other aspects of the organization such as human factors, culture, strategy and objectives. Organizations should be prepared for the broad spectrum of impacts when introducing new technologies, for instance the multiple costs associated and the social implications.

To make a correct selection, managers must carefully analyze the different technological alternatives. The objective of this assessment is to evaluate the consequences of introducing a technology. If the assessment is done before the technology adoption and investment, it reduces the risk of ineffective investment decisions [3]. When an organization integrates new technologies without being aware and prepared for its impact, it can lead to serious problems.

Technology Assessment in Software Development Projects Using a System Dynamics Approach: A
Case of Application Frameworks

117

In the context of technology management, technology assessment (TA) can be defined as "a systematic attempt to foresee the consequences of introducing a particular technology in all spheres it is likely to interact with" [1], which can include the inside of an organization, project or process as well as the society. Because the technology has been not implemented yet, or may not even exist, the consequences have not occurred at the time of the assessment. In the technology assessment field, forecasting does not provide an assertion of the future, but a visualization of what it might be and the opportunities it may bring.

There are three basic forecasting methods: extrapolation, expert opinion and modeling. The extrapolation method involves obtaining historical data, fit a curve to them and extend the values into the future. When an expert opinion method is employed, experts in certain field try to reach a consensus on the likely future. Modeling consists in using mathematical formulas to represent the relationships that exist between variables in the real world.

One of the modeling approaches used in technology assessment is in a system dynamics. System dynamics is a "perspective and set of conceptual tools that enable us to understand the structure and dynamics of complex systems" [4].

In the software industry, developments are made in complex and dynamic systems involving several interrelated elements such as people, processes, methods, products, technologies, tools and techniques [5]. These interrelations create a feedback system where a change or improvement in one area creates effects in others parts of the system directly or indirectly. Additionally, system complexity might stem from its risks and uncertainties, dynamic behavior or changes over time [6].

Simulation techniques like system dynamics can be used to understand these interrelations and to analyze in advance the impacts of changes or improvements such as new technology adoption. In this chapter we describe the use of system dynamics models and simulations to assess technologies in the context of software engineering.

This chapter is organized as follows; in the first section we review the field of technology assessment and explain an intermediate level assessment using system dynamics. In the next section, the field of software process simulation modeling and a general overview of system dynamics are presented. The subsequent sections are composed of studies applied to technology adoption and a study regarding a reuse technology. At the end of the chapter conclusions and future research of this study can be found.

2. Technology assessment

2.1. Background

The field of technology assessment started in the 1950s with studies that attempted to forecast technological trends to help government agencies and large corporations in their technological investments; later, in the public decision domain of the United States, technology assessment started to focus on studying all the effects of technologies still to come [7].

There have been different approaches in time and in the public and private sector regarding technology assessment. However, there is a common concern about the current and future developments of technology and to improve the alignment between technological and societal developments, which includes the activities of corporations.

2.2. Application from a technology manager's perspective

The use of technologies is one of the main components that determine the success of an organization. From the technology manager's perspective, technology can be defined as "the ways and means by which humans produce purposeful material artefacts and effects" [1], this definition includes the material elements, such as hardware and means, but also the soft elements, such as: knowledge, ways and methods.

To take the best advantage from technology, a firm needs to analyze the technological alternatives and their consequences thoroughly. These consequences must considered not only in the short term but as far as possible and contemplate its implications to the full context, for the firm and the environment in which it operates. Technology assessment can provide the framework to create this type of analysis to evaluate the firm's products, processes and supporting technologies current and future positions.

The assessment of new technologies can be considered from two perspectives: the introduction of a new technology to improve a production process or the launch of a technology into the market. The former can refer to a process innovation, such as the development of new production equipment, new production processes and the introduction of new technologies in a process. The second, known as product innovation, refers to the development of new products for the consumer.

According to [8], technology assessment can provide managers information to:

Support decision making for:	Help the firm:
R&D direction	Corporate strategic technology planning
New technology adoption	Perform value chain analysis
Incremental improvement in existing technologies	Identify market advantage
Level of technology friendliness	Avoid being preempted in the marketplace
'Make -or-buy' decisions	Stay on the cutting edge
Optimal expenditure of capital equipment funds	Define the cutting edge
Market diversification	Maximize the use of information
	Minimize product makespan
	Achieve a competitive advantage (in cost, quality, or time to market)

Table 1. Examples of technology assessment application areas in the organization. [8]

A basic methodology for technology assessments can be outlined in the five steps described in Figure 1:

Technology Assessment in Software Development Projects Using a System Dynamics Approach: A
Case of Application Frameworks

119

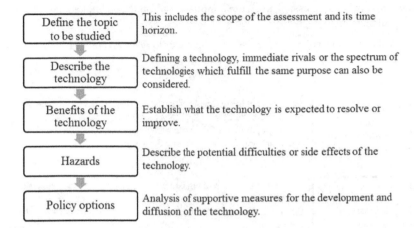

Define the topic to be studied	This includes the scope of the assessment and its time horizon.
Describe the technology	Defining a technology, immediate rivals or the spectrum of technologies which fulfill the same purpose can also be considered.
Benefits of the technology	Establish what the technology is expected to resolve or improve.
Hazards	Describe the potential difficulties or side effects of the technology.
Policy options	Analysis of supportive measures for the development and diffusion of the technology.

Figure 1. Diagram of the five basic steps in a technology assessment methodology

A crucial element in a technology assessment is the gathering of information relevant to the issue; it involves discussions and dialogue between multiple actors in its development. The information can reside in a variety of sources, including opinions, attitudes, fears, interests.

The possibility of using simulation to assess technologies has been emphasized in different works, in [9], it is asserted that performing simulations in organizations can help evaluate technology strategies; in [10], the same approach of using process simulation modeling to evaluate issues of technology and tool adoption is used.

One of the techniques that can be used in a technology assessment is system dynamics. It provides tools which can be used to include the multiple perspectives involved in the decision and analyze them. Using this technique, it is possible to adopt a holistic view of the company to understand it as a system, considering its products, processes and supporting technologies and enable the development of an intermediate level assessment of a technology [11].

2.3. Intermediate level assessment

Dynamic simulation models and maps of the anticipated domain of application of a technology make it possible to evaluate its impact at different levels of the organization, for example strategic and project levels. These simulations will assist decision makers and it can reduce their decision risks.

An intermediate level assessment of a technology can be created using a system dynamics model; it can capture the domain-centered orientation of low level approach but provide a balanced assessment of impact across the whole domain of application. This assessment will be located between a specific or low-level of a detailed analysis of the technology characteristics and functionalities, and a high level view provided by an economic analysis. In this intermediate level it is necessary to create a model based on the technology and the domain where it will be used.

In [11], Wolstenholme suggests a three step methodology to create this intermediate assessment:

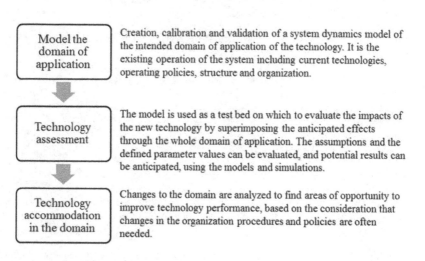

Figure 2. Diagram of the three step methodology to create an intermediate assessment.

Some adjustments that might be projected in the simulation are: changes in the organizational boundary, responsibilities of activities, elimination of delay, increasing or reducing capacities, information paths and policies, or reengineering processes.

These stages of the methodology proposed by Wolstenholme are related to the basic technology assessment methodology described earlier in the chapter. The definitions of the technologies, areas of application, expected benefits and difficulties, as well as the current and future policies, can serve as input in the creation of the models. At the same time, the knowledge acquired with the model simulations feedback the technology assessment.

3. Modeling and simulation in software

3.1. Software process simulation modeling

From the perspective of software research, the area focused on simulation and modeling of software processes and projects is Software Process Simulation Modeling (SPSM). SPSM can be helpful in the definition of why to use simulations, what to model and simulate, which simulation method to use, how to perform this activities and which parameters to use; it provides recommendations when using and implementing this type of analysis in an organization, as well as describe potential issues.

When performing a technology assessment on software engineering technologies, SPSM provides relevant information to execute the assessment, because it focuses purely on issues of software development and maintenance.

Concerning the why, Kellner et al. [6] identified and grouped the purposes for using simulations into six categories: strategic management, planning, control and operational management, understanding, training and learning, process improvement and technology adoption.

Process improvement and technology adoption can be grouped together in the same category because both support decisions for improvements (such as the priorization of various proposals) through forecasting of potential impacts; the difference is that the first explores the improvement of processes, and the second deals with the introduction of new technologies. These purposes can also be related to planning in the sense that they are concerned with the future, but the main difference to planning is that the objective is not to make a plan but to analyze an improvement effort.

Within the field of software process simulations there are various approaches to simulate software processes and projects. In [12], ten simulation paradigms were identified; however, system dynamics is the most widely used technique in SPSM.

Using system dynamics modeling and simulations, it is possible to [9]:

• Research about a great variety of aspects of a software process at a macro or micro levels.

• Evaluate different alternatives and make the selection based on their implications.

• Compare life cycle processes, defect detection techniques, testing strategies and different methodologies and tools.

• Represent systems and processes as they are currently implemented, or as-is, but also as planned or expected in the future, or to-be.

Two of the cornerstones of SPSM with system dynamics are the works of [9] and [13]. The difference between these two works is where each establishes the boundary of their view; [9] integrates processes outside the project boundary, [13] does not. The work in [9], as most of the work in the area, considers a process perspective. A software process can be defined as "a set of activities, methods, practices and transformations that people use to develop and maintain software and the associated products (e.g., project plans, design documents, code, test cases, and user manuals)" [14].

In order to determine what will be simulated, it is necessary to have the basic purpose for the simulation model. A simulation model is a computerized model used to represent a dynamic system. When a simulation model is being developed, it is necessary to identify the main elements of the process under study, interrelations and behaviors to abstract the process. Along with the specific issues to be analyzed, the scope of the model, the input parameters and the resulting variables and an abstraction of the process will also have to be defined. A diagram of these interrelations between the aforementioned elements can be seen in Figure 3.

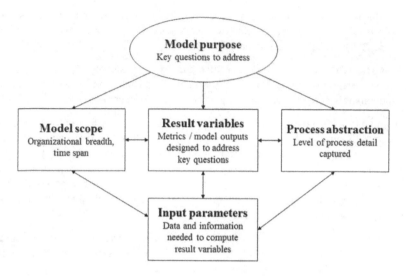

Figure 3. Diagram showing the interrelationships between model elements [15].

Usually, the scope of the simulation is among the following: a portion of the life cycle, a development project, multiple concurrent projects, long term product evolution and long term organization. It is necessary to define the timespan, which can be under twelve months, between 12 and 24 months and over 24 months, and organizational breadth, which can be less than a product/project team, one product/project team or multiple product/project teams.

The input parameters as well as their measure units are determined by the desired result variables and the identified process abstractions. These variables can be viewed to remain constant for the duration of the simulation, or they may vary over time. The variables in the model may be independent or have interdependencies, which can also be represented in the model. The result variables, as indicated by their name, are the variables we want to obtain from the simulation.

A series of parameters have been identified from the literature. Table 2 lists the typical elements included in process abstraction, input parameters and result variables.

3.2. Difficulties of using simulation models in software engineering

Although there are numerous benefits of using simulation models in software engineering, there are also difficulties. These difficulties are not specific to assessments in technology adoption, they can occur in all applications.

Multiple articles mention that a great amount and detail of data from the technology development processes is needed to implement a SPSM. For technology adoption, this is of particular importance because improvements are analyzed through changes in the initial parameters. In [10], it is considered that to be able to make a software adoption assessment, a CMMi maturity level 5 is needed.

Other difficulty derives from the specialized knowledge required to use the simulation approaches. For example, it is necessary to understand the simulation approach and the simulation software package. A firm will have to invest time and resources to instruct people inside the company or integrate the experience from the outside.

Process abstraction	Input parameters	Result variables
key activities and tasks;	amount of incoming work;	effort/cost;
primary objects;	effort for design as a function of size;	cycle time;
vital resources;	defect detection efficiency during	defect level;
activity dependencies, flows of	testing and inspections;	staffing requirements over time;
objects among activities and	effort to code rework as a function of	staff utilization rate;
sequencing;	size and number of defects identified	cost/benefit, return of investment or
iteration loops, feedback loops and	for correction;	other economic measures;
decision points;	defect removal and injection rates	productivity;
other structural interdependencies.	during code rework;	queue lengths.
	decision point outcomes; number of	
	rework cycles;	
	hiring rate staff; staff turnover rate;	
	personnel capability and motivation,	
	over time;	
	amount and effect of training provided;	
	resource constraints;	
	frequency of product version releases	

Table 2. Examples of typical elements included in process abstraction, input parameters and result variables. [6]

4. System dynamics

4.1. Background

The field of system dynamics was first developed by Jay Forrester in 1950. It is grounded in the theory of nonlinear dynamics and feedback control developed in mathematics, physics, and engineering. Under this approach, a model is constructed using a defined nomenclature and principles; after that, the model is simulated using a software program.

System dynamics provides a methodology that allows integrating multiple perspectives when analyzing complex and dynamic systems [16]. To create, understand and use of system dynamics models and simulations, it is necessary to be familiar with the principles upon which the models are constructed. System dynamics utilizes a continuous approach with feedback mechanisms.

From the system perspective, a system can be classified depending on its perspective of change over time and its past actions' effects.

If the variables that describe the system change over time the system it's consider dynamic, otherwise it is static. The dynamic behavior is part of the complexities found in real life systems. Depending on how the variables change over time, a dynamic system it can be divided in continuous, discrete or combined. If the system variables change over time without breaks or irregularities, it is continuous, compared with discrete where change is presented instantaneously at separated points in time. In a continuous system, the events are not individually tracked, instead they are considered as an aggregate that can be described using differential equations. Under this approach, time changes at a constant rate.

Another principle in System Dynamics is the closed or feedback system approach. In a closed system, the future behavior of the system is influenced by its past actions, compared to an open perspective where the outputs are not influenced by the inputs. These feedback loops are what generate and control all that changes thru time [9, 17].

System dynamics does not predict the future *per se*; the accumulations in the system and its structure are what produce the system behavior [17].

4.2. System dynamics models and diagrams

To use system dynamics, a model representing the problem must be developed. A model is an abstraction of a real system [6]. In a model, it is not necessary to include all the system's elements but only those elements relevant to the problem under analysis.

The models can be used to experiments different scenarios by changing information, parameters and structure. It is possible to test 'what if' or 'tradeoff' experiment without affecting the real system because of its cost or because it is impossible to do so without altering it. These experiments allow determining how real and proposed systems perform in time [9].

In system dynamics there are different models and diagrams to describe the boundary of a model and its structure. These models and diagrams are:

- Model boundary chart - Used to list the key variables that are considered endogenous for the model (and are therefore included) and which are considered exogenous (therefore being excluded).

- Subsystem diagram - They present the major subsystems and their coupling, the boundary of the model and the level of aggregation by presenting the different organizations or agents.

- Causal loop diagram (or "causal diagram") - It is used to diagram the feedback structure of a system. It is composed by causal links among variables with arrows to describe cause and effect. One of its limitations is that it lacks the representation of the stock and flow structure necessary to keep track of system accumulations.

- Stock and flow maps - Are used to describe underlying physical structure of a system. Stocks keep track of the accumulations in terms of material, money and information.

Flows are the rates of increase of decrease in stocks. The stocks are an important part of system dynamics because they characterize the state of the system. This diagram is especially important because it is the one which is simulated by the simulation software package.

Stock and flow maps are created using the following representation:

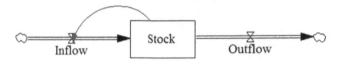

Figure 4. Basic system dynamics stock and flow map

In Table 3, each of the basic elements represented in stock and flow map are described:

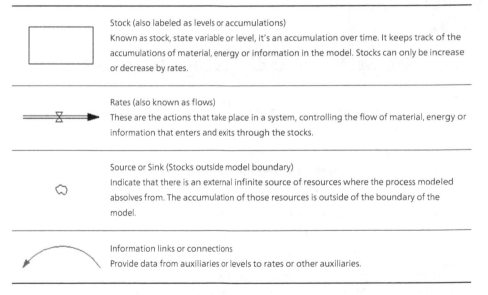

	Stock (also labeled as levels or accumulations) Known as stock, state variable or level, it's an accumulation over time. It keeps track of the accumulations of material, energy or information in the model. Stocks can only be increase or decrease by rates.
	Rates (also known as flows) These are the actions that take place in a system, controlling the flow of material, energy or information that enters and exits through the stocks.
	Source or Sink (Stocks outside model boundary) Indicate that there is an external infinite source of resources where the process modeled absolves from. The accumulation of those resources is outside of the boundary of the model.
	Information links or connections Provide data from auxiliaries or levels to rates or other auxiliaries.

Table 3. Basic elements represented in stock and flow map

Similarly to technology assessment, system dynamics uses several sources of information, such as: numerical data, written description and observations from mental models.

System dynamics models can be simulated using one of the different simulation software packages such as STELLA® [18], iThink® [19] and Vensim® [20].

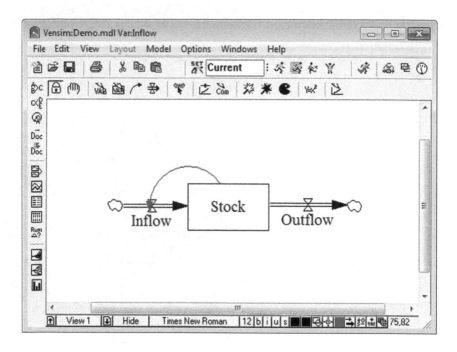

Figure 5. Screenshot from Vensim® simulation software.

5. System dynamics in the assessment of technology

There are previous examples where system dynamics has been applied in studies to evaluate the effects of technology adoption in software processes:

Fourth Generation Languages	In [21] we can find a study researching the dynamic effects of reuse and fourth generation languages in a rapid application development using a system dynamics model. The model consists of four stages: requirements, design, coding and approval phases. Different relationships between efforts are believed to exist within different programming languages. The learning curve is also deemed to vary across languages as do error rates. In the model, the reuse reduces the development effort. The study considers that models can help a project manager plan a strategy by revealing the effects of using various types of programming and their reuse levels in relation to the effort.

Commercial-of-the-Shelf components	In [22], a study analyzes the interrelation between glue code development and Commercial-of-the-Shelf components (COTS) integration.
	In a component based development, development of new code is needed to integrate components into the system; this is called glue code, glueware or binding code. This code is often specific to the project, and it could bring technical risks. It is difficult to decide when a software development team should start the development of glue code and integrate it into the system. One of the benefits of using the model in [22] is that it can aid in the decision of when to begin the development of glue code and integration.
	The model is formed by four main sections: the COTS glue code development, the COTS component factor, human resources and application development / integration.
	As a general rule, they found that glue code should be developed at the end of the development of the application, and the integration should occur at the beginning of the glue code development.
	In [23], the implications of specific features of a COTS-based software development process are analyzed.

Table 4. System dynamics applied to technology adoption in software development

Examples of simulation models applied to assess technologies, outside software development, include:

Defense organization	Wolstenholme, 2003: [11]
	The assessments of two management information systems in the defense industry are documented.
Pharmaceutical	Wolstenholme, 2003: [11]
	A case of a new anesthetic drug in a hospital is described.
Automotive	Yearn Min Kim, 2007: [24]
	Presents a forecast of the demand for in car navigation in the automotive industry; the forecast considers elements such as potential adopters, vehicle demand, navigation price, speed camera and speed fine.

Table 5. System dynamics applied to technology adoption in other fields

6. Case study

In this section, a system dynamics simulation model created to assess a reuse technology is described.

6.1. Application frameworks

Application frameworks is an object-oriented reuse technique [25] generally targeted to a particular application domain [16]. A framework is a "semi-complete application that can be specialized to produce custom applications" [26].

At early stages of the object-oriented field, an object was considered an appropriate abstraction level to achieve reusability, but objects were a very specialized for a particular application. Later, the concept of application framework appeared, these frameworks are more adequate to a achieve reusability in an object-oriented development [27].

Frameworks are a subsystem design composed by a collection of abstract and concrete classes with an interface between them. A framework is not a complete application by itself; it needs to be extended to create a subsystem or more specific application by instantiating specific plug-ins [27], it can be compared to a tool box [28].

Their objective is the reuse of larger-grain components and high-level designs which differ to other reuse models such as libraries [29], design patterns and components [30].

6.1.1. Framework benefits

The benefits of a framework-based development are: savings in time to market [24, 28], effort and costs, along with improvements in productivity and quality enhancement [28, 30]. These improvements are achieved as an obvious consequence of reusing code [24, 28], but also because reusing the designs that are part of the framework creates uniformity in the application [25]. Reuse of components can provide improvements in developer productivity and improve software quality, performance, reliability and interoperability [31].

According to [28], a framework "should allow you to develop the application quickly and easily and should result in a superior finished application".

6.1.2. Framework difficulties

Application frameworks also have negative sides. According to [27], application frameworks are a good approach for taking advantage of reuse but they have a high cost when they are introduced in software development processes and projects. Additionally, they create a dependency on the developed applications [25], which can affect the software maintenance.

An application framework is itself a complex software, and in order to use it, it's first necessary to understand it, which consumes software engineers' time, the learning curve can even take months [27]. This learning curve can be a decisive factor in making a framework technology profitable [28]. It is important to estimate if implementing the technology in a project or organization would be a viable alternative.

6.2. Problem statement

Application framework technologies provide benefits but add difficulties to development projects. The main benefits described in literature arise from reuse, while the main difficulty is the learning curve; in this chapter we explore the effects of the latter. As mentioned before, framework learning can be a decisive factor in making this technology profitable in organizations and projects.

6.3. Application area

The presented model can be used to analyze the adoption of an application framework technology in software development projects.

6.4. Model overview

Given that an application framework allows the reuse of code and patterns via code, and main benefits and difficulties of the framework technology occur during the software coding phase, it was defined that this phase was the domain of the technology. The model is focused on the use of the technology, not on its creation. Elements from other models described in system dynamics literature and main elements in framework-based software development were included.

The model is composed of four interrelated subsections: software development, human resources, projects and framework learning.

The human resource section, presented in Figure 6, was modeled using a subsection from the model in [13]. This section allows simulating different scenarios for active developers in a project. Two categories of developers are considered: beginners and experts. The first category takes a predefined period of time to turn into expert. Both categories have a potential productive capacity.

There are two main issues affecting potential productive capacity. The first is that the expert programmers will be asked by beginners to help resolve questions and problems, absorbing productive capacity. Furthermore, a loss in productivity is considered due to communication between team members. The more team members, the more time is consumed in communication exchanges.

Given that in the literature it is exposed that becoming proficient at a framework technology can take up to several projects, the model includes a section called Projects. This section allows simulating the development of more than one consecutive projects. A new project will begin as soon as the previous one is completed, until the set number of projects has been completed. This section is presented in Figure 7.

Because the model is presented by segments, the relationship between the "In progress" rate from the project section and the "Project artifacts" rate from the Software development section was omitted.

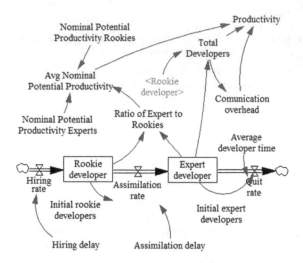

Figure 6. Human resource subsection

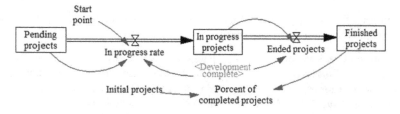

Figure 7. Projects section

The software development section, shown in Figure 8, derives from different models where software development is simulated. In the model, the term artifact is used to not limit the measure unit to be simulated, for example: function points, object oriented function points, code lines, etc. Similar to the model in [13], a productivity multiplier is included as a result of learning from the project. The incidence of defects that will have to be rebuilt and might affect the conclusion of the project is also taken into account.

The framework learning subsection represents the learning of the technology. As mentioned before, framework learning is a decisive factor to consider when introducing this technology. For this reason, this subsection will be described in detail.

The learning curve is an element which frequently appears to be a difficulty in the implementation of new technologies in software development projects. The role played by this element is fundamental. Among other consequences, it can affect programmer productivity

Technology Assessment in Software Development Projects Using a System Dynamics Approach: A
Case of Application Frameworks

131

and increase difficulty in maintenance. Since the learning curve varies over time in a project, it can be analyzed in detail using simulations in the assessment of a technology.

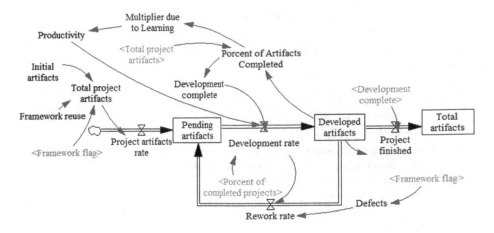

Figure 8. Software development section

The model considers that the programmer can learn about the framework while developing artifacts in a project. This knowledge is lost if the programmer quits the project. The knowledge left over after a programmer leaves depends on the amount of programmers working in the project. If the sole programmer in a project leaves, all knowledge is lost. If only one programmer in a bigger team leaves, one part of the knowledge is lost, while other parts have been transferred and distributed to other team members.

In the model, it is also considered that programmers begin with no framework knowledge. Nonetheless, this section enables to run simulations where developers start with varying knowledge levels due to education or training. The framework learning subsection described before is presented in Figure 9.

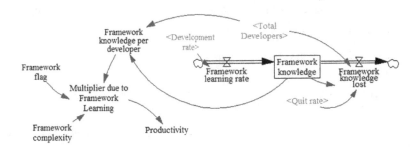

Figure 9. Framework learning subsection

The framework learning subsection is composed by four main elements:

Framework knowledge	In the model, the programmer gains framework knowledge by developing artifacts. Framework knowledge is the accumulation of artifacts developed by the programmers. It is the sum of all the framework knowledge of the developers.
Framework knowledge per developer	This variable is used to calculate the average knowledge per developer.
Framework complexity	The framework complexity variable is used to indicate how many artifacts a programmer must develop to reach the maximum expected benefits in productivity as a result of the framework usage.
Multiplier due to Framework Learning	This variable is the application framework learning curve. This learning curve generates an improvement or decline in developer productivity as a result of framework knowledge. The position in the learning curve is calculated based on framework complexity and the number of artifacts built with the framework per developer.

Table 6. Main elements framework learning subsection

6.5. Simulation model calibration and results

This section presents the main data used to configure the model and the results obtained from the simulations. The software package Vensim® was selected to generate the simulations.

The research data in [29] was used as a base to configure the model simulations. In [29], the authors made an exploratory case study where one subject developed five applications using a framework and four applications without one. These nine applications were developed in a research and development company in a six month period using a proprietary framework. The findings show that the productivity and quality is increased in both stances, however, using a framework to develop applications increased productivity more significantly.

In the present study, the model will use the following configuration settings:

• The simulated artifacts will be based on Object Oriented Function Points (OOFP).

• The simulation will comprise five projects

• The initial number of expert programmers is set to one and the amount of beginner programmers is cero.

• The average project size will be 1370 OOFP. After a decrease of 80% due to framework reuse, the average project size will be 274 OOFP.

• Nominal potential productivity of expert programmers will be established at 1 OOFP/ Hour.

• A 10% defect percentage will be considered.

Technology Assessment in Software Development Projects Using a System Dynamics Approach: A
Case of Application Frameworks

133

- To reach the plateau of the learning curve, the subject must develop 1370 OOFP using the framework.

The data presented above remained unchanged in the presented simulations with the purpose of identifying the impacts or effects of an isolated element, the learning curve. In this study the impacts of two different application framework learning curves are explored. In scenario 1, the curve is generated from the data in [29]. In the cited study, productivity was not found to decrease in the technology learning period; instead, during the five projects, productivity was shown to increase. As a result, productivity in scenario 1 is not lowered by the learning and use of the framework.

A similar curve is used in scenario 2, but productivity is lower until 20% of framework knowledge is achieved. Once the 20% knowledge has been reached, the negative effect in productivity is overcome, but productivity is still lower than in scenario 1 until framework knowledge reaches 42%. From this point forward, both curves are equal. Figure 10 exhibits the two learning curves (Scenario 1 and Scenario 2) used in the model. The curves are represented based on their effect in programmer productivity.

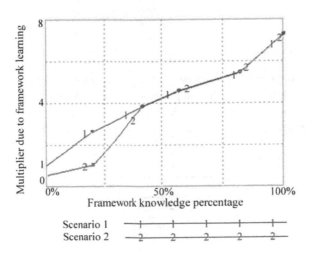

Figure 10. Learning curves used in the model.

Two simulations, one for each scenario described before, were executed. The results of the simulations are presented in Figure 11 and Table 7. Figure 11 was generated using the Vensim® report functionalities.

In Figure 11, the amount of OOFP developed in time is presented. Each of the peaks in the blue line (Scenario 1) or the red line (Scenario 2) corresponds to the conclusion of each simulated project. The x-axis represents time measured in hours; as it increases, the amount of developed artifacts in each project is increased in the y-axis. Once all artifacts, in this case

OOFP, in a project have been developed, the amount of developed artifacts is reset to cero to begin a new project.

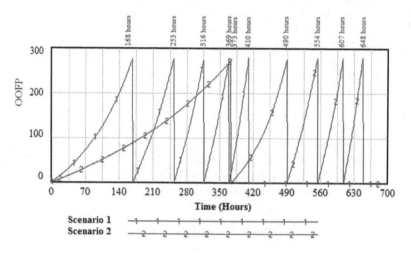

Figure 11. Results of the simulation – Developed artifacts

Table 7 lists the hour in which each of the five projects was concluded in sequence by scenario. In scenario 1, the five simulated projects end after 410 hours, while in scenario 2 they end after 648 hours, a difference of 238 hours. The data show a 58% percent increase in scenario 2 compared to scenario 1. In other words, projects in scenario 2 would end 58% later than in scenario 1.

Project number	Time of conclusion	
	Scenario 1	Scenario 2
1	168 hours	373 hours
2	253 hours	490 hours
3	316 hours	554 hours
4	369 hours	607 hours
5	410 hours	648 hours

Table 7. Time of conclusion for each simulated project

Results show that the learning curve has an important effect in development time because of its relationship to productivity. Considering that learning is achieved through project development, this impact in productivity can consequently inhibit framework learning.

Technology Assessment in Software Development Projects Using a System Dynamics Approach: A
Case of Application Frameworks

135

7. Conclusions

7.1. General conclusions

The accelerated rhythm of technology advancements pressures managers and organizations to evaluate what is the best alternative for their particular needs. Deciding which technologies to implement in software projects and processes is a challenging task, but of paramount importance.

The area of technology assessment provides decision makers with guidelines to achieve this task; it relies on different techniques to analyze the impact of these technologies and the aspects involved. Simulation modeling techniques, such as system dynamics, make it possible to analyze the impact of adopting a technology before its implementation.

System dynamics has been applied in the assessment of new technology adoption in the field of software development. It offers the unique advantage of being able to include a variety of perspectives and information sources, including qualitative and quantitative data. This technique also allows isolating particular elements for the purpose of impact analysis, as in the case study presented. This is significantly beneficial in the study of complex system with multiple interrelated elements and can provide meaningful insights.

The area of model simulation in software has a broad range of applications, from support of strategic decisions to project planning and also, as described in this chapter, technology adoption.

7.2. Case conclusions

A model to evaluate the use of frameworks in software development projects was presented in the chapter. The learning curve is an important aspect to consider in the development and use of frameworks because it can make it unviable to use a technology in a project or organization. For this reason, this element was selected to be isolated and studied using system dynamic simulations.

Two scenarios with different learning curves were simulated to analyze the effect of framework learning. The changes in the learning curve had a significant effect in the time of conclusion of a series of sequential projects. We consider that the results of the simulations presented concur with the literature, which is a positive finding, given that the model aims to represent as best as possible the reality of using framework technologies in projects.

8. Future research

Various research opportunities can be developed from this study; the structure of the model presented for application frameworks will continue to be researched, experimenting with different configurations and initial parameters. The importance of the learning curve for this technology will be analyzed, through the use of more model simulations. As for the relation between technology assessment and software process simulation models, it is possible to make a systematic literature review from the perspective of technology adoption.

Author details

José Ignacio Muñoz Hernández[1*], José Ramón Otegui Olaso[2] and Alejandro Gutiérrez López[2]

*Address all correspondence to: JoseIgnacio.Munoz@uclm.es

1 University of Castilla-La Mancha UCLM, Spain

2 University of the Basque Country UPV/EHU, Spain

References

[1] Ernest B. Technology in Context: Technology Assessment for Managers. 1st ed. London: Taylor & Francis; 1998.

[2] Torkkeli M, Tuominen M. The contribution of technology selection to core competencies. Int J Prod Econ 2002 6/11;77(3):271-284.

[3] Daim TU, Intarode N. A framework for technology assessment: Case of a Thai building material manufacturer. Energy for Sustainable Development 2009 12;13(4): 280-286. http://www.sciencedirect.com/science/article/pii/S0973082609000751 (accessed 6 August 2011).

[4] Sterman JD. Business Dynamics: Systems Thinking and Modeling for a Complex World. United States of America: McGraw-Hill; 2000.

[5] Pfahl D, Lebsanft K. Using simulation to analyse the impact of software requirement volatility on project performance. Information and Software Technology 2000;42(14): 1001-1008. http://linkinghub.elsevier.com/retrieve/pii/S095058490000152X (accessed 11 November 2010).

[6] Kellner MI, Madachy RJ, Raffo DM. Software process simulation modeling: Why? What? How? J Syst Software 1999;46(2-3):91-105. http://linkinghub.elsevier.com/retrieve/pii/S0164121299000035 (accessed 13 June 2011).

[7] Van Den Ende J, Mulder K, Knot M, Moors E, Vergragt P. Traditional and Modern Technology Assessment: Toward a Toolkit. Technological Forecasting and Social Change 1998 0;58(1–2):5-21. http://linkinghub.elsevier.com/retrieve/pii/S0040162597000528 (accessed 3 August 2012).

[8]   Henriksen ADP. A technology assessment primer for management of technology. Int
      J Technol Manage 1997;13(5/6):615-638. http://inderscience.metapress.com/content/
      yu9yq058feyd6avh/ (accessed 12 August 2012).

[9]   Madachy RJ. Software Process Dynamics. 1st ed.: Wiley-IEEE Press; 2008.

[10] Raffo DM, Wakeland W. Moving Up the CMMI Capability and Maturity Levels Us-
      ing Simulation. Pittsburgh: Software Engineering Insitute, Carnegie Mellon Universi-
      ty; 2008. http://www.sei.cmu.edu/library/abstracts/reports/08tr002.cfm (accessed 3
      September 2012).

[11] Wolstenholme EF. The use of system dynamics as a tool for intermediate level tech-
      nology evaluation: three case studies. J Eng Technol Manage 2003;20(3):193. http://
      linkinghub.elsevier.com/retrieve/pii/S0923474803000183 (accessed 21 October 2010).

[12] Zhang H, Kitchenham B, Pfahl D. Software process simulation modeling: Facts,
      trends, and directions. Proceedings of 15th Asia-Pacific Software Engineering Con-
      ference, APSEC 08, 2-5 December 2008, Beijing, China.

[13] Abdel-Hamid T, Madnick S. Software Project Dynamics: An Integrated Approach.
      Prentice Hall; 1991.

[14] Paulk MC, Weber CV, Garcia SM, Chrissis MB, Bush M. Key Practices of the Capabil-
      ity Maturity Model Version 1.1. Software Engineering Institute 1993;Paper 171.
      http://www.sei.cmu.edu/reports/93tr025.pdf (accessed 3 September 2012).

[15] Münch J, Armbrust O, Kowalczyk M, Soto M. Software Process Simulation. In:
      Münch J, Armbrust O, Kowalczyk M, Soto M (eds.) Software Process Definition and
      Management. Berlin Heidelberg, Springer; 2012. p187-210. Available from http://
      www.springerlink.com/content/xv1p58082762075w/ (accessed 3 September 2012).

[16] Fayad ME, Schmidt DC, Johnson R. Implementing Application Frameworks: Object-
      Oriented Frameworks at Work. John Wiley & Sons; 1999.

[17] PhD seminar in system dynamics conducted by Jay W. Forrester at the MIT Sloan
      School of Management [DVD]. Albany, NY1: Sloan School of Management. System
      Dynamics Group; 1999.

[18] isee systems, inc. STELLA Systems Thinking for Education and Research. http://
      www.iseesystems.com/softwares/Education/StellaSoftware.aspx (accessed 3 Septem-
      ber 2012).

[19] isee systems, inc. iThink Systems Thinking for Business. http://www.iseesys-
      tems.com/softwares/Business/ithinkSoftware.aspx (accessed 3 September 2012).

[20] Ventana Systems, Inc. Vensim Software. http://www.vensim.com/software.html (ac-
      cessed 3 September 2012)

[21] Lo KW. Reuse and high level languages. Center for Systems and Software Engineer-
      ing, University of Southern California 1999; CS599 Final Report. http://
      sunset.usc.edu/classes/cs599_99/projects/reuse.pdf (accessed 3 September 2012).

[22] Kim WK, Baik J. Dynamic model for COTS glue code development and COTS integration. Center for Systems and Software Engineering, University of Southern California 1999; CS599 Final Report. http://sunset.usc.edu/classes/cs599_99/projects/COTS.pdf (accessed 3 September 2012).

[23] Ruiz M, Ramos I, Toro M. Using Dynamic Modeling and Simulation to Improve the COTS Software Process. Lecture Notes in Computer Science 2004;3009:568-581. http://www.springerlink.com/index/MDEF683VQ3LKHGNB.pdf (accessed 12 September 2011).

[24] Kim YM. A system dynamics model for the technological forecasting of automotive in-car navigation market. Proceedings of Second Intl Forum on Strategic Technology, IFOST 2007, 3-6 October 2007, Ulaanbaatar, Mongolia.

[25] Johnson RE. Frameworks = (components + patterns). Commun ACM 1997;40(10): 39-42. http://dl.acm.org/citation.cfm?doid=262793.262799 (accessed 25 October 2011).

[26] Johnson RE, Foote B. Designing Reusable Classes. J Object-Oriented Programming 1988;1(2):22-35.

[27] Sommerville I. Ingeniería de Software. 7th ed. Madrid: Pearson Educación S.A.; 2005.

[28] Nash M. Java Frameworks and Components: Accelerate Your Web Application Development. Cambridge University Press; 2003.

[29] Morisio M, Romano D, Stamelos I. Quality, Productivity, and Learning in Framework-Based Development: An Exploratory Case Study. IEEE Trans.Softw.Eng. 2002;28(9):876-888. http://ieeexplore.ieee.org/lpdocs/epic03/wrapper.htm?arnumber=1033227 (accessed 12 September 2011).

[30] Polancic G, Horvat RV, Rozman I. Improving Object-Oriented Frameworks by Considering the Characteristics of Constituent Elements. J Inf Sci Eng 2009;25(4): 1067-1085. http://www.iis.sinica.edu.tw/page/jise/2009/200907_07.pdf (accessed 12 September 2011).

[31] Fayad M, Schmidt DC. Object-oriented application frameworks. Commun ACM 1997 October;40(10):32-38. http://dl.acm.org/citation.cfm?doid=262793.262798 (accessed 22 June 2011).

Chapter 7

Technical Performance Based Earned Value as a Management Tool for Engineering Projects

José Ignacio Muñoz Hernández,
José Ramón Otegui Olaso and Julen Rubio Gómez

Additional information is available at the end of the chapter

1. Introduction

In the project management, one of the keys for the project success is the control of the project in terms of performance, progress and cost. The first practice to control a project is just checking the financial reporting and therefore the project control is only based on the cost control (budgeted versus actual incurred cost). But the experience, especially from projects that failed, shows that a more efficient project management requires controlling also the project performance and scheduling besides the cost.

The Earned Value Management (EVM) is a methodology that integrates the management of project scope, schedule and cost. It has been widely and successfully used for over 40 years and it should apply to any project, large or small, at any industry. The interest about the EVM has increased in the last ten years and a substantial amount of research has been carried out regarding the EVM, generating new extensions of the methodology and specific applications. In parallel, many guidelines and rules have been published to implement properly the EVM, coming mainly from the U.S. Government who originally developed the EVM, but not only, because also the ANSI organization has established a standard defining the EVM.

The contribution of this work is twofold. First of all, it aims to identify the principal lines of research in the project management control using earned value management across the academic research and organizations practice. Following, the attention is focused on the characteristics of one of these research lines, the PBEV. This extension pretends to be not only an EVM parameters enhancement but a complete managing system based on the technical performance, which could be a useful EVM improvement to be used in engines engineering projects where the technical objectives are the main target. Secondly, it lies in the intention of this work to evaluate the applicability and the efficiency of the PBEV with two case studies of two real-life engineering projects of combustion engines development for energy generation applications.

2. Earned value management

2.1. EVM standard

Basically, the EVM requires a fixed point-of-reference, given by the project baseline schedule and the budget at completion (BAC), in order to periodically measure the project performance along the life of the project. Project performance, both in terms of time and costs, is determined by comparing the three key parameters of the EVM, planned value (PV), actual costs (AC) and earned value (EV), resulting in performance variances known as schedule variance (SV=EV-PV) and cost variance (CV=EV-AC) and performance indexes as the schedule performance index (SPI=EV/PV) and the cost performance index (CPI=EV/AC). See [4] and [5]. Figure 1 shows the EVM key parameters.

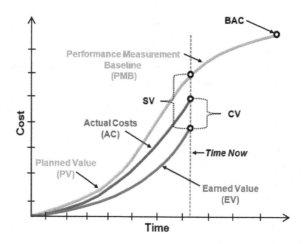

Figure 1. EVM key parameters.

For a better understanding of the state of the art of the EVM it is necessary to know its origins, widely commented in [1] and [2], and the official sources of EVM guidelines and standards.

The earned value concept originally came from industrial engineers in factories in the early 1900s who for years have employed a three-dimensional approach to assess true "cost-performance" efficiencies. To assess their cost performance, they compared their earned standards (the physical factory output) against the actual cost incurred. Then, they compared their earned standards to the original planned standards (the physical work they planned to accomplish) to assess the schedule results. These efforts provided earned value in its most basic form. Most important, the industrial engineers defined a cost variance as the difference between the actual costs spent and the earned standards in the factory. This definition of a cost variance is perhaps the indication to determine whether one uses the earned-value concept. Later on, in 1965 the United States Air force acquisition managers defined 35 criteria which capture the essence of earned value management. Two years later the U.S. Department of Defense (DoD) adopted these same criteria as part of their Cost/Schedule Control Systems Criteria (C/SCSC). Then, in 1996, after a rewrite of the C/SCSC 35 criteria by private industry, the DoD accepted the rewording of this criteria under a new title called Earned Value Management System (EVMS), and the total number of criteria was reduced to 32. In 1998, National Defense Industrial Association (NDIA) obtained acceptance of the Earned Value Management System in the form of the American National Standards Institute, termed the ANSI/EIA-748 Standard, see [3]. From this on, many research lines have been developed trying to improve the EVM, and they are commented in the following point.

2.2. EVM further research lines

As mentioned in the introduction, the use of EVM and parts of it, or tailoring it to specific situations continue to grow and a substantial amount of research is being carried out. The main investigations could be grouped at least in six big research lines as follows.

1. EVM and fuzzy determination of EV. One of the major difficulties in the determination of EV is the evaluation of in-process work. There are some techniques to quantify the accomplished work and some of them propose fuzzy techniques. As explained in [6] and [7].

2. EVM forecast accuracy. Some studies about EVM are focusing on the improvement of the accuracy of EVM to calculate forecasts of project cost and schedule at completion. See references [8], [9] and [10]

3. EVM and Earned Schedule. Another object of studies is an enhancement to Earned Value Management, called Earned Schedule that appeared in 2003. This method propose the parameters of the EVM in terms of time instead of cost, and it could be postulated to give more forecasting accuracy in the latest stages of the project. See [11]. The author of this research line criticized the use of the classic SV and SPI metrics since they give false and unreliable time forecasts near the end of the project. Instead, he provided a time-based measure to overcome this unreliable behavior of the SV and SPI indicators. This earned schedule method relies on similar principles than the earned value method, but translates the monetary metrics into a time dimension. He reformulated the time performance measures SV and SPI to SV(t) and SPI(t) (where the t between brackets is used to distinguish with the original performance measures) and has shown that they have a reliable behavior along the whole life of the project. Since its introduction, a considerable stream of publications on the project time performance measurement and/or the earned schedule concept has been published in the academic literature. Statistical validation of the time performance indicators and stability studies on empirical data can be found in [12].

4.   EVM to integrate risk management. There are project tools for risk analysis mainly focused on network based techniques, such as, CPM (Critical Path Method) and PERT (Program Evaluation and Review Technique) and others based in statistical probabilities like Montecarlo Simulations. Nevertheless, EVM contains no guidelines on risk management. Therefore, some investigations are conducting to integrate risk analysis in the EVM. First, including risk management activities in the WBS (Work Breakdown Structure) and second considering that Estimated at Completion (EAC) should be based on both project performance and quantified risk assessments. Other investigations try to combine the risk analysis tool results, for example Montecarlo simulations, with EVM metrics. See [13].

5.   EVM to integrate quality. The quality of the product must be tracked in an engineering project to evaluate the success or failure of the product. Some studies are proposing to apply a quality factor to the EV, such as, reducing the EV when poor quality occurs. See [14].

6.   EVM to integrate technical performance. In engineering projects not only cost and schedule must be tracked but also the technical performance. Further investigations are being done to integrate technical requirements accomplishment with EVM. One proposal is an EVM extension called Performed-Based Earned Value (PBEV). See [15-18].

This last extension of the EVM integrating the technical performance, the PBEV, will be deeply analyzed in the following points due to it is not only a new manner of measuring the earned value but it is an entire management system which include concepts and procedures to manage the technical issues that could enhance the engineering projects in general. Therefore, it will be studied its suitability to engine engineering projects.

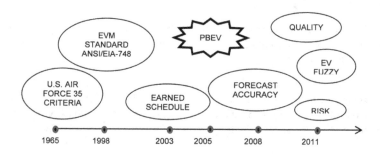

Figure 2. EVM evolution and research lines.

3. PBEV: The EVM extension focusing on technical performance

3.1. PBEV introduction

The Performance Based Earned Value (PBEV) is an enhancement to the EVM standard that overcomes the standard's shortcomings with regard to measuring technical performance and quality because it is based on standards and models for systems engineering, software engineering, and project management. The distinguishing feature of PBEV is its focus on the customer requirements. PBEV provides principles and guidance for cost effective processes that specify the most effective measures of cost, schedule, and product quality performance.

The PBEV has been formulated by the author Paul J. Solomon, see [16], whom experience at programs management and concretely with the EMVS explains the origin of the PBEV. Solomon manages the EVM for Northrop Grumman Corporation, Integrated Systems and he has participated in programs for the U.S. Government as the B-2 Bomber, Global Hawk and F-35 Joint Strike Fighter. He is on the board of the National Defense Industrial Association, Program Management Systems Subcommittee that authored ANSI/EIA-748.

3.2. EVM shortcomings

The cancellation of the US Navy's A-12 Avenger stealth aircraft program in January 1991 resulted in research during the 1990s, which investigated the reliability of EVM cost prediction and the project management failures in accurately reporting project performance, using U.S. Department of Defense (DOD) project data. These research findings have come to be regarded as generally applicable across all project types using EVM across multiple industry sectors. The U.S. Government Accountability Office (GAO), for example, studied failures in acquisition of weapons systems and information technology systems. Some report of the GAO concluded that DOD paid billions in award and incentive fees regardless the contractors not held cost goals, schedule goals and the program did not capture early on the warnings needed to effectively manage program risk. The GAO concluded that, if EVM is not properly implemented, the data may be inaccurate and misleading.

The EVM standard, which is formulated in the ANSI/EIA Standard-748-A-1998, has significant shortcomings with regard to measuring technical performance and quality. First, the EVMS standard states that EV is a measurement of the quantity of work accomplished and that the quality and technical content of work performed are controlled by other processes. A program manager should ensure that EV is also a measurement of the product quality and technical maturity of the evolving work products instead of just the quantity of work accomplished. Second, the EVMS principles address only the project work scope. EVMS ignores the product scope and product requirements. Third, EVM is perceived to be a risk management tool. However, EVMS was not designed to manage risk and does not even mention the subject.

Thus, the EVM will provide more reliable information for analysis and decision-making if the EVM guidelines are augmented by guidance regarding maintaining the technical baseline, measuring technical performance, and managing risk.

3.3. U.S. DOD policy and industrial standards

Most of the guidelines and best practices for planning and measuring the technical perform-ance considering also the EVM are provided by two sources: the U.S. Department of Defense (DOD) and some industrial standards.

First of this sources, the U.S. DOD, issued an acquisition policy which states that programs implement Systems Engineering Plans (SEP). DOD also published guides to implement the policy, such as:

1. The Defense Acquisition Guidebook (DAG).

2. The Systems Engineering Plan Preparation Guide (SEPPG).

3. The Work Breakdown Structure Handbook (MIL-HDBK-881A (WBS)).

4. The Integrated Master Plan (IMP).

5. The Integrated Master Schedule (IMS) Preparation and Use Guide.

6. The DOD guides refer also to EVMS. See [19].

These guides show the enormous importance that the U.S. DOD gives to measuring technical performance, hence, according to U.S. DOD it should be one of the parameters to control during project progress. See [20] and [21].

The second source of guidelines for planning and measuring the technical performance are some industry and professional standards, that have been incorporating the technical performance accomplishment for the project control and they have been at the same time sources themselves for defining the DOD policy, as well as:

1. The Institute of Electrical and Electronics Engineers (IEEE) 1220.

2. EIA 632.

3. "A Guide to the Project Management Body of Knowledge" (PMBOK). See [22] and [23].

4. Capability Maturity Model Integration (CMMI). See [24].

In the following points are explained the main concepts defined by the U.S. DOD guidelines and the industrial standards that are used by the PBEV. Figure 3 shows the guidelines and standards and their sources for technical performance treatment at projects.

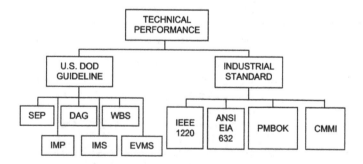

Figure 3. Guidelines and standards to manage technical performance.

3.4. PBEV guidelines

The PBEV is a set of principles and guidelines that specify the most effective measures of cost, schedule, and product quality performance. It has several characteristics that distinguish it from standard EVMS:

- Plan is driven by product quality requirements, not work requirements.

- Focuses on technical maturity and quality, in addition to work.

- Focuses on progress toward meeting success criteria of technical reviews.

- Adheres to standards and models for SE, software engineering, and project management.

- Provides smart work package planning.

- Enables insightful variance analysis.

- Ensures a lean and cost-effective approach.

- Enables scalable scope and complexity depending on risk.

- Integrates risk management activities with the performance measurement baseline.

- Integrates risk management outcomes with the Estimate at Completion.

PBEV augments EVMS with 4 additional principles and 16 guidelines. The following are PBEV principles that set it apart from EVMS:

1. Product scope and quality. Integrate product scope and quality requirements into the performance measurement baseline. This principle focuses on customer satisfaction, which is based on delivery of a product that meets its quality requirements and is within the cost and schedule objectives.

2. Product quality requirements. Specify performance toward satisfying product quality requirements as a base measure of earned value. A product quality requirement is a characteristic of a product that is mandatory in order for the product to meet verified customer needs.

3. Risk management integration. Integrate risk management with EVM.

4. Tailor PBEV. Tailor the application of PBEV according to the risk.

And the PBEV guidelines are listed following.

1. Establish product quality requirements and allocate these to product components.

2. Maintain bidirectional traceability of product and product component quality requirements among the project plans, work packages, planning packages, and work products.

3. Identify changes that need to be made to the project plans, work packages, planning packages, and work products resulting from changes to the products quality requirements.

4.   Define the information need and objective to measure progress toward satisfying product quality requirements.

5.   Specify work products and performance-based measures of progress for satisfying product quality requirements as base measures of earned value.

6.   Specify operational definitions for the base measures of earned value, stated in precise, unambiguous terms that address communication and repeatability.

7.   Identify event-based success criteria for technical reviews that include development maturity to date and the products's ability to satisfy product quality requirements.

8.   Establish time-phased planned values for measures of progress toward meeting product quality requirements, dates of frequency for checking progress, and dates when full conformance will be met.

9.   Allocate budget in discrete work packages to measures of progress toward meeting product quality requirements.

10.   Compare the amount of planned budget and the amount of budget earned for achieving progress toward meeting product quality requirements.

11.   Use Level of Effort method to plan work that is measurable, but is not a measure of progress toward satisfying product quality requirements, final cost objectives, or final schedule objectives

12.   Perform more effective variance analysis by segregating discrete effort from Level of Effort.

13.   Identify changes that need to be made to the project plans, work packages, planning packages, and work products resulting from responses to risks.

14.   Develop revised estimates of costs at completion based on risk quantification.

15.   Apply PBEV coverage to the whole work breakdown structure or just to the higher risk components.

16.   Apply PBEV throughout the whole system development life cycle or initiate after requirements development.

3.5. PBEV concepts

The PBEV feeds on concepts from the U.S. DOD guidelines and industrial standards mentioned above. In the following points are explained in detail the main concepts used in the PBEV coming from those sources, such as, the Systems Engineering Plan (SEP), the products metrics and quality, the success criteria, the Technical Performance Measurements (TPM) and Capability Maturity Model Integration (CMMI). The concepts that feed the PBEV are shown in figure 4.

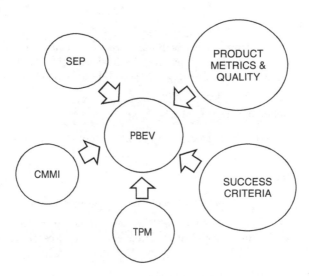

Figure 4. The concepts that feeds the PBEV

1.   Systems Engineering Plan (SEP)

The purpose of the Systems Engineering Plan (SEP) is to help programs develop their Systems Engineering (SE) approach, providing a firm and well-documented technical foundation for the program. The SEP is a living document in which periodic updates capture the program's current status and evolving SE implementation and its relationship with the overall program management effort. Although the detailed content of each SEP is customized according to the particulars of a program and each update may vary depending on the program's acquisition phase, using a common framework encourages sound technical planning throughout the program's life cycle. The emphasis should be on the rigor of the technical planning as captured in the SEP, not on the SEP itself. The SEP also serves as a common reference to achieve shared stakeholder insight regarding a program's planned technical approach. It provides a documented understanding of how the program will accommodate cost, schedule, performance, and sustainment trades; the expected products of the SE effort; and how these products will contribute to program decision making.

- Products metrics and quality

The IEEE 1220 and the EIA 632 have similar guidance regarding product metrics and quality. Product metrics allow assessment of the product's ability to satisfy requirements and to evaluate the evolving product quality against planned or expected values. Of equal importance are a disciplined requirements traceability process and a requirements traceability database.

- Success criteria

The standards discuss the importance of holding technical reviews at various stages of development to assure that all success criteria have been met. IEEE 1220 provides success criteria to be used at major technical reviews. For example some of the success criteria for a preliminary design review are the following:

i.   Prior completion of subsystem reviews.

ii.   Determine whether total system approach to detailed design satisfies the system baseline.

iii.   Unacceptable risks are mitigated.

iv.   Issues for all subsystems, products, and life-cycle processes are resolved.

The success criteria should be defined in a SEP or other technical plan. The customer should review this plan with the supplier and reach agreement on the success criteria to be used at technical reviews.

- Technical Performance Measurement (TPM)

Technical Performance Measurement (TPM) are defined and evaluated to assess how well a system is achieving its performance requirements. TPM uses actual or predicted values from engineering measurements, test, experiments, or prototypes. IEEE 1220, EIA 632 and "A Guide to the Project Management Body of Knowledge" (PMBOK guide) provide similar guidance for TPM planning and measurement and for integrating TPM and EVM.

- Capability Maturity Model Integration (CMMI)

The CMMI provides many practices that augment the EVMS guidelines. CMMI also lists Typical Work Products (TWPs) within process areas. To ensure traceability of product quality requirements to work tasks and work products, these TWPs, or similar artifacts, should be the outcome of work packages. Here are some TWPs in CMMI.

TWPs include the following:

i.   Product-component requirements.

ii.   Activities diagrams and use cases.

iii.   Verification criteria used to ensure requirements have been achieved.

iv.   Exit and entry criteria for work products.

3.6. PBEV pros and cons

Following are listed the advantages and disadvantages of the PBEV that it has been used also in the following case studies analyzed during this work.

The main PBEV pros are:

• It integrates effectively the technical performance with the cost and schedule control.

• It collects the best practices of the U.S. DOD guides and industry and professional standards related with project management and then apply the EVM.

• It is based on client or product requirements.

By the other hand, the PBEV could have the following cons:

• It gives a very large methodology to implement to be able to integrate the technical performance in the project management control but there is no mention to any parameter, or metric definition to implement easily that allows calculating EVM variances or indexes with results of technical performance status.

• It only influences somehow the work breakdown structure but not the EVM metrics.

• The customer should review the SE plan with the supplier and reach agreement on the success criteria to be used at technical reviews. It would be more objective to define a technical parameter to evaluate success criteria, as for example, the strength safety margin of a mechanical part.

4. PBEV model adaptation to industrial production environment

They have been performed two case studies of the applicability and the efficiency of the EVM standard and one of its extensions, the PBEV which is based on technical performance, for the engines engineering projects control in the energy field.

4.1. Type of projects analyzed

They have been analyzed two real-life projects of engine engineering in the energy field performed between 2004 and 2005 by a company dedicated to the development and manufacturing of combustion engines for power generation.

These projects are characterized by including the typical phases of the industrial engineering projects, such as, the concept design phase, detailed design phase, simulation, prototyping, testing and launch mass production, where the level of compliance of the technical objectives is strongly important besides the cost and scheduling control. The two projects analyzed are also featured by the fact that they are large projects in the energy field as they have budgets of several million euros, durations between 2-4 years and the risk of developing new products for the market. The large size of the projects makes the breakdown structure of the tasks (WBS) to be also quite large. Another property that characterizes these two projects, and in general all the engine development projects in the energy field, is that they are pretty similar to engine development projects in the automotive field since the product is similar. Therefore, the engines made for energy field are fed with many of the technologies used in automotive field, as well as, with the methodologies and requirements to create new products that meet the market requirements. The competitiveness for obtain improved technical results makes that in both, energy projects and automotive projects, the technical objectives are a critical parameter to control for the project, just with the costs and the schedule. Hence, the PBEV could be a suitable and useful tool for integrating scope, schedule, cost and technical performance in these projects. Moreover, in both fields, energy and automotive, the continuous technology evolution makes the technical objectives to be overpassed from one project to the following. For that reason is a real-life practice in these fields that at new product development projects, with the existing technology in that moment, that, if the technical objectives are not met but they are inside of an allowable range, the project could be accepted with penalties.

In summary, the projects analyzed with PBEV are characterized for being classical industrial engineering projects with large size of the WBS, also for having the technical performance as a key parameter to control where the technical objective could be in a range of tolerance for accept the successful of the project. The project number 1 consists on the development of a new engine of higher efficiency and power than the existing ones in the company and the project number 2 was dedicated to development a new injection system for the current engines. Figure 5 shows the type of engine developed in project 1 and 2.

Figure 5. Type of projects analysed: engine engineering projects

4.2. Project information

The project 1 began in January 2004 and was delayed one year by material procurement problems but once they were negotiated new deadlines with suppliers in 2005 was launched again and began an overall tracking of the technical targets, cost and schedule. In this project, the engineering consultancy delivered a design that was slightly below the technical targets of engine performance and efficiency, which was penalized in his fees. The engine efficiency target was 42% but an engine with 40% was obtained. Finally, the project was completed in January 2008, bringing to the market the new engine with a very good acceptance and overall rating of satisfactory.

Regarding the project 2, it began also in 2005 and although the design and material procurement were on time, the project turned 180° in 2007 when it was found that to accomplish the technical targets of the new injection system it was necessary to implement a technology 10 times more expensive than originally planned. This event made management direction to stop the project after making a rough estimation of the overall cost of the project with the implementation of the new technology.

From these two project data, here is proposed to apply the PBEV to analyze if it is possible to report the status of the project in time, cost and technical compliance, to predict future states.

In the two projects analyzed the available information during the duration of all the project was the following documents and reports.

a.   A project specification document which collects all the technical specifications to meet with contractual nature.

b.   A Statement Of Work (SOW) document which collects the definition of all the tasks to be performed in the project and has contractual nature.

c.   The initial baseline schedule in an MS project file. It is detailed with several tasks levels and including the starting and finishing dates, as well as, the dependences between them.

d.   The total budget of the project made at the beginning of the project and broken down in general spending issues. The general spending issues are the following six: material costs, the tooling investments, engineering hours, outsourcing expert consultancy support hours, testing costs, and trips. Intermediate budget estimation was not performed during the project progress.

e.   The general accounting of the project with the invoiced costs per month. They are available the figures of monthly costs of the general issues.

f.   A monthly report with the technical, economical and scheduling tracing. The technical part of the report included a list of all manufacturing drawings and the testing results. Figure 6 shows engine control units screen where technical parameters are monitored, as for example, the engine performance. These technical values are the reference to check the technical accomplishment. The economical part of the report included cost monthly figures of the general spending issues collected from the invoices. The scheduling part of the report included approximate deadlines for the critical tasks but not a detailed scheduling track.

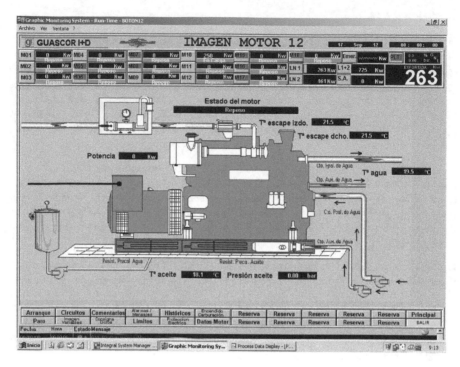

Figure 6. Engine control unit monitoring engine performance

4.3. Methodology

Due to fact that the available data in the two projects analyzed are the initial budget, accumulated costs and the engineering hours accumulated only for the higher level of the WBS, the EVM is applied only at this top level. This way of controlling is called top-down project control. According to reference [13], the EVM offers calculation methods yielding reliable results on higher WBS levels, which greatly simplify final duration and completion date forecasting. These early warning signals, if analyzed properly, define the need to eventually go down into lower WBS levels when action thresholds are exceeded. In conjunction with the project schedule, it allows taking corrective actions on those activities which are in trouble (especially those tasks which are on the critical path).

The methodology used to implement the EVM based on technical performance in the two projects under study has two steps. The first step is to apply the EVM standard and check if the results are consistent. The second step is to apply the PBEV through the use of penalties in the obtained EV when the technical requirements are not met and check if the results are consistent with the reality of the project. An objectively way of establish the penalties depends on the technical parameter selected to control and on the technical performance measurement, therefore, it depends on the internal mechanisms of each company to evaluate the objectives compliance as explained in reference [16].

To carry on the first step, this is to apply the EVM standard on the two projects, from the available data, one begins by determining the values of the three basic parameters of the EVM, and this is, PV, AC and EV. Because PV is not available monthly, it is considered the hypothesis of distributing uniformly the initial budget between the months of the project, and then the PV accumulated is lineal. The figures of the monthly AC are available in the project data. Finally, to calculate the EV, the engineering hours performed are taken as the indicator of work accomplished. And from the 3 basic parameters they are calculated the Cost and Schedule Variances (CV and SV) and the Cost and Schedule Performance Indices (CPI and SPI) to check if the results of de EVM are consistent.

To implement the second step, this is to calculate the PBEV, a reduction of the EV is applied in the monthly work accomplished when the technical objectives are not met, so in this way it is shown that the project is farther from the objectives and then from finish than planned.

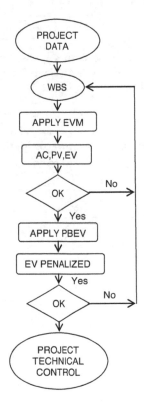

Figure 7. Methodology to apply PBEV to engine engineering projects.

5. Case studies: Applying PBEV to engine engineering projects

5.1. Research questions

The two case studies carried out with engine engineering projects have been raised as a series of research questions to be answered in order to draw conclusions. The research questions are the following:

- Q1: Is it possible to apply the EVM standard to engines engineering projects only with the data that is usually controlled in such projects?

- Q2: Is it possible to apply the PBEV to engines engineering projects?

- Q3: Does the PBEV capture the real project progress in engines engineering projects with only the project information available?

- Q4: Is it useful the information resulting from the PBEV to take decisions in the engine engineering projects?

5.2. Results

In this point are summarized the two case studies results. Table 2 shows the project 1 cumulative cost data. The Budget At Completion (BAC) for project 1 was 2.427.950 euros. The costs and BAC are used as input to calculate the EVM and the PBEV parameters which are shown in table 2.

PROJECT 1 - CUMULATIVE COST DATA (€)					
ITEM	Oct-07	Jun-08	Aug-08	Nov-08	Dec-08
MATERIALS	416.686	444.229	445.059	471.645	472.736
INVESTMENT	279.356	279.356	279.356	279.356	279.356
LABOR HOURS	308.880	400.650	424.890	474.750	487.560
TESTING HOURS	2.935	4.036	4.353	4.688	4.688
ENGINEERING OUTSOURCING	1.443.296	1.522.324	1.522.605	1.526.015	1.526.015
TRIPS	13.021	16.435	16.435	22.524	22.730
TOTAL	2.464.174	2.667.030	2.692.698	2.778.978	2.793.085

Table 1. Project 1 cumulative cost data in euros.

PROJECT 1 - EVM AND PBEV CALCULATED PARAMETERS

PARAMETER	Oct-07	Jun-08	Aug-08	Nov-08	Dec-08
PV	1.824.282	2.169.235	2.255.473	2.384.831	2.427.950
AC	2.464.174	2.667.030	2.692.698	2.778.978	2.793.085
EV	1.538.160	1.995.156	2.115.866	2.364.159	2.427.950
CV	-926.014	-671.874	-576.832	-414.819	-365.135
SV	-286.122	-174.079	-139.607	-20.672	0
CPI	0,62	0,75	0,79	0,85	0,87
SPI	0,84	0,92	0,94	0,99	1,00
PBEV	1.300.000	1.600.000	1.700.000	1.900.000	2.000.000

Table 2. Project 1 EVM and PBEV calculated parameters.

Figure 8 shows the calculated EVM parameters and figure 9 shows the calculated PBEV against the EVM for project 1.

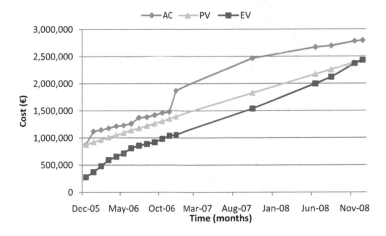

Figure 8. Project 1 calculated EVM parameters AC, PV, EV.

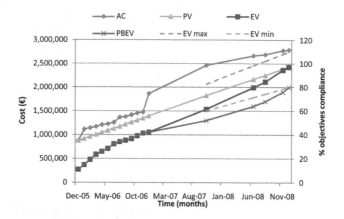

Figure 9. Project 1 EVM versus PBEV.

Following is explained the calculation of the points of figure 9. Let instant time June 2008, for example, the Actual Cost, AC, is 2.667.030€, which is a project data defined in table 1. The Planned Value, PV, is the proportional part of the total budget, BAC, corresponding to the month 30, that is June 2008, over the total number of months of the project which is 36 from December 2005 to December 2008. Previously, to the initial BAC of the 2004 which was a project input data of 2.427.950€ is subtracted the costs from 2004 to 2005 of 875.661€ to consider the initial BAC updated to 2005 because in this year the project was launched again after a delay an all the project data start in 2005. The Earned Value, EV, is calculated with the rate labor hours at June 2008 over the total labor hours multiply by the BAC, resulting in 1.995.156 €. For calculating the PBEV it is considered a penalty of 25% in the EV because the engine efficiency target of 42% is not achieved and instead of that is obtained 40%.

Table 3 shows the project 2 cumulative cost data. The Budget At Completion (BAC) for project 2 was 2.616.831 euros. The costs and BAC are used as input to calculate the EVM and the PBEV parameters which are shown in table 4.

PROJECT 2 - CUMULATIVE COST DATA (€)					
ITEM	Jan-07	Feb-07	Mar-07	Apr-07	May-07
MATERIALS	297.018	298.606	315.487	315.742	324.575
INVESTMENT	54.467	54.467	54.467	54.467	54.467
LABOR HOURS	307.170	320.940	338.220	346.530	366.480
TESTING HOURS	111.854	126.781	138.642	147.510	159.630
ENGINEERING OUTSOURCING	139.290	139.290	139.290	139.290	139.290
TRIPS	18.770	18.770	18.770	18.770	19.370
FUEL	2.822	3.000	3.256	3.408	3.986
TOTAL	931.391	961.854	1.008.132	1.025.717	1.067.798

Table 3. Project 2 cumulative cost data in euros.

PROJECT 2 - EVM AND PBEV CALCULATED PARAMETERS					
PARAMETER	Jan-07	Feb-07	Mar-07	Apr-07	May-07
PV	2.030.091	2.176.776	2.323.461	2.470.146	2.616.831
AC	931.391	961.854	1.008.132	1.025.717	1.067.798
EV	307.170	320.940	338.220	346.530	366.480
CV	-624.221	-640.914	-669.912	-679.187	-701.318
SV	-1.722.921	-1.855.836	-1.985.241	-2.123.616	-2.250.351
CPI	0,33	0,33	0,34	0,34	0,34
SPI	0,15	0,15	0,15	0,14	0,14
PBEV	300.000	300.000	300.000	300.000	300.000

Table 4. Project 2 EVM and PBEV calculated parameters.

Figure 10 shows the calculated EVM parameters and figure 11 shows the calculated PBEV against the EVM for project 2.

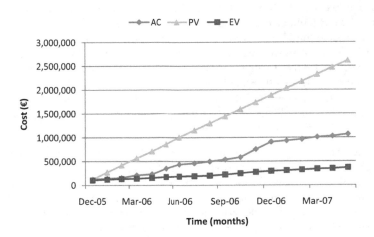

Figure 10. Project 2 calculated EVM parameters AC, PV, EV.

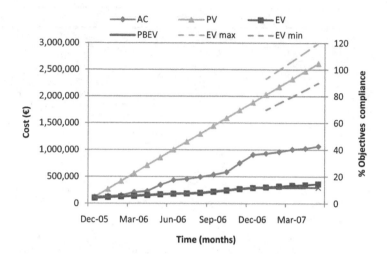

Figure 11. Project 2 EVM versus PBEV.

Now, it is possible to give answers to the proposed questions.

- Q1: The results presented in figures 8 and 10 show that it is possible to apply the EVM to the higher level of the WBS in projects where the only information is the monthly accounting and the work progress, so the EVM parameters could be obtained and a project control with EVM is possible. For checking that the obtained values for AC, PV and EV are consistent with the EVM, it could be checked that in general when a project is finished satisfactorily the SV is zero and the SPI is 1, due to the fact that at the end of the project the earned value will match with the planned value. For the project 1, which finished satisfactorily, at the end of the project the EV fitted with the PV and therefore the SV is zero and the SPI is 1, as can be seen in table 2. By the other hand, for the project 2, which was intentionally terminated by the managing director, the EV does not reach the PV and thus, the SV never becomes zero and the SPI does not tend to 1. The EVM application in both projects is consistent with EVM rules and with the reality of the projects.

- Q2: The result of PBEV presented in figures 9 and 11 shows that, once the EVM is applied, is also applicable the PBEV if they are defined penalties for when the technical targets are not fully accomplished. The manner to quantify the penalties depends on each company's internal procedure and his technical performance measurements, as it is mentioned in [16]. For the project 1, after applying the PBEV criteria, which means a reduction in the EV value because in this project was established an engine efficiency of 42% and it reached only to 40%, the obtained PBEV is still inside an admissible range of objectives accomplishment and thus it is accepted the engine with performance 40% to be released. In the case of the project 2, the application of the PBEV is also possible but it has not much influence because of the project circumstance of have been stopped by Managing Director. Thus, this project had not time to correct the situation even the PBEV gives a more realistic status of objectives accomplishment.

- Q3: As the PBEV using the EV with penalty adjust the real situation of the project integrating technical targets, cost and schedule, it captures properly the real situation in an engine development project where the accomplishment of the technical targets is a key parameter.

- Q4: The PBEV gives useful information for the project control and helps to the decision making process regarding the technical targets compliance besides the cost and scheduling control. The proposed EV with penalty in the PBEV reduce the EV near the end of the project and this implies to take a decision between accepting the project finish with the obtained technical results under the objectives or the necessity of continue the project until the objectives are achieved.

6. Conclusions

The following conclusions could be extracted from the present work.

- The EVM method to project control has been widely extended in the last ten years and a substantial amount of research has been generated in academic literature and industrial guidelines to evaluate the method and proposing extensions to improve it.

- The research regarding the EVM could be classified in six main research lines according with this work. The research lines try to enhance the EVM adding to it concepts, such as, the product quality, the project risk analysis, the technical performance, a fuzzy determination of the EV, or analysing for example the EVM forecast accuracy, or even, there is a research line, the Earned Schedule, that proposes another point of view for the EVM in terms of time instead of costs.

- One of the most complete extensions of the EVM is the PBEV, that is not only a new EVM metric to evaluate the technical performance but a complete system of guidelines for project management that integrates the technical performance measurement in all the project management procedures including the EVM.

- The PBEV could be applied to industrial projects such as engine engineering projects in the energy field and it could be applied at the top level of the project WBS.

- The case study number 1 shows that PBEV allows capturing the real project status including the technical performance. In this case, the PBEV reduce the EV near the end of the project and this implies to take a decision between accepting the project finish with the obtained technical results under the objectives or the necessity of continue the project until the reach the objectives.

- The case study number 2 shows that PBEV could be applied to a project that finish dramatically but it does not contribute significantly to give early warnings of the health of the project if the project finished by other issue regardless the technical performance.

7. Future research

In a complementary way to the PBEV that considers EV with penalty when the technical target is not achieved, it could be proposed other penalties in the PBEV variables, such as, the penalty to the AC to capture the reduction in supplier's fees when the technical targets are not met or when the project suffers an unexpected increase in the cost of any task or component. This AC with penalty could be an improvement in the project management as a tool also for risk management.

The future research in general will include probably new metrics in the EVM methodology to take into account issues like risk analysis or quality and technical performance for a more efficient project control. By the moment, these issues depend on the practitioners, to be included in the project scheduling by means of milestones which is quite subjective.

Moreover, the combination of other project management techniques with EVM is also a good practice that is under developing by some authors.

Which it is sure is that the EVM will take place as a common tool in the project management as Gantt did in the past because it is easy, quick, an efficient.

Author details

José Ignacio Muñoz Hernández[1], José Ramón Otegui Olaso[2] and Julen Rubio Gómez[3]

*Address all correspondence to: JoseIgnacio.Munoz@uclm.es

1 University of Castilla-La Mancha UCLM, Spain

2 University of the Basque Country UPV/EHU, Spain

3 Dresser-Rand Inc., U.S.A.

References

[1] Fleming Q, Koppelman J. Earned Value Project Management. The Journal of Defense Software Engineering 1998.

[2] Kwak Y, Anbari F. History, Practices, and Future of Earned Value Management in Government: Perspectives from NASA. Project Management Journal 2012;43(1) 77-90.

[3] American National Standards Institute. Earned Value Management Systems. ANSI/EIA-748-A-1998. 1998.

[4] Anbari F. Earned Value Project Management Method and Extensions. Project Management Journal 2003;34(4) 12-23.

[5]   Fleming Q, Koppelman J. Start with "Simple" Earned Value…on All Your Projects. The Measurable News 2006.

[6]   Moslemi L, Shadrokh S, Salehipour A. A fuzzy approach for the earned value management. International Journal of Project Management 2011;29(6) 764-772.

[7]   Noori S, Bagherpour M, Zorriasatine F. Designing a control mechanism for production planning problems by means of earned value analysis. Journal of Applied Sciences 2008;8(18) 3221-3227.

[8]   Vanhoucke M, Vandevoorde S. Earned Value Forecast Accuracy and Activity Criticality. The Measurable News, 2008; 13-16.

[9]   Fleming Q, Koppelman J. The Two Most Useful Earned Value Metrics: the CPI and the TCPI. The Journal of Defense Software Engineering 2009.

[10] Pajares J, López-Paredes A. An extension of the EVM analysis for project monitoring: The Cost Control Index and the Schedule Control Index. International Journal of Project Management 2011;29(5) 615-621.

[11] Lipke W. Shedule Is Different. The Measurable News 2003; 31-34.

[12] Lipke W, Zwikael O, Henderson K, Anbari F. Prediction of Project Outcome: The application of Statistical Methods to Earned Value Management and Earned Schedule Performance Indexes. International Journal of Project Management 2009;27(4) 400-407.

[13] Vanhoucke M. Measuring the Efficiency of Project Control Using Ficticious and Empirical Project Data. International Journal of Project Management 2011; To Appear(2) 252-263.

[14] Xu J. Project Integrated Management Based on Quality Earned Value. Proceedings of the 2nd International Conference on Information Science and Engineering 2010; 432-435.

[15] Solomon P. Practical Software Measurement, Performance-Based Earned Value. The Journal of Defense Software Engineering 2001.

[16] Solomon P. Performance-Based Earned Value. The Journal of Defense Software Engineering 2005.

[17] Solomon P. Practical Performance-Based Earned Value. The Journal of Defense Software Engineering 2006.

[18] Solomon P. Integrating Systems Engineering with Earned Value Management. The Measurable News 2008; 4.

[19] DOD Earned Value Management Implementation Guide. U.S. Department of Defense. http://www.acq.osd.mil/evm/resources/guidance-references.shtml (accessed 10 February 2012).

[20] Humphreys and Associates. http://www.humphreys-assoc.com/evms/evms-implementation-ta-a-18.html (accessed 17 February 2012).

[21] National Aeronautics and Space Administration. NASA. http://evm.nasa.gov (accessed 3 February 2012).

[22] PMBOK, 2004. A guide to the Project Management Body of knowledge. Project Management Institute; 2004.

[23] PMI, 2005. The Practice Standard for Earned Value Management. Project Management Institute; 2005.

[24] Kerzner H. Project Management. A Systems Approach to Planning, Scheduling and Controlling. John Wiley and sons 2003.

Chapter 8

Modeling and Linear Programming in Engineering Management

William P. Fox and Fausto P. Garcia

Additional information is available at the end of the chapter

1. Introduction

Consider planning the shipment of needed items from the warehouses where they are manufactured and stored to the distribution centers where they are needed.

There are three warehouses at different cities: Detroit, Pittsburgh and Buffalo. They have 250, 130 and 235 tons of paper accordingly. There are four publishers in Boston, New York, Chicago and Indianapolis. They ordered 75, 230, 240 and 70 tons of paper to publish new books. There are the following costs in dollars of transportation of one ton of paper:

From \ To	Boston (BS)	New York (NY)	Chicago (CH)	Indianapolis (IN)
Detroit (DT)	15	20	16	21
Pittsburgh (PT)	25	13	5	11
Buffalo (BF)	15	15	7	17

Management wants you to minimize the shipping costs while meeting demand. This problem involves the allocation of resources and can be modeled as a linear programming problem as we will discuss.

In engineering management the ability to optimize results in a constrained environment is crucial to success. Additionally, the ability to perform critical sensitivity analysis, or "what if analysis" is extremely important for decision making. Consider starting a new diet which needs to healthy. You go to a nutritionist that gives you lots of information on foods. They recommend sticking to six different foods: Bread, Milk, Cheese, Fish, Potato and Yogurt: and provides you a table of information including the average cost of the items:

	Bread	Milk	Cheese	Potato	Fish	Yogurt
Cost, $	2.0	3.5	8.0	1.5	11.0	1.0
Protein, g	4.0	8.0	7.0	1.3	8.0	9.2
Fat, g	1.0	5.0	9.0	0.1	7.0	1.0
Carbohydrates, g	15.0	11.7	0.4	22.6	0.0	17.0
Calories, Cal	90	120	106	97	130	180

We go to a nutritionist and she recommends that our diet contains not less than 150 calories, not more than 10g of protein, not less than 10g of carbohydrates and not less than 8g of fat. Also, we decide that our diet should have minimal cost. In addition we conclude that our diet should include at least 0.5g of fish and not more than 1 cups of milk. Again this is an allocation of recourses problem where we want the optimal diet at minimum cost. We have six unknown variables that define weight of the food. There is a lower bound for Fish as 0.5 g. There is an upper bound for Milk as 1 cup. To model and solve this problem, we can use linear programming.

Modern linear programming was the result of a research project undertaken by the US Department of Air Force under the title of Project SCOOP (Scientific Computation of Optimum Programs). As the number of fronts in the Second World War increased, it became more and more difficult to coordinate troop supplies effectively. Mathematicians looked for ways to use the new computers being developed to perform calculations quickly. One of the SCOOP team members, George Dantzig, developed the simplex algorithm for solving simultaneous linear programming problems. The simplex method has several advantageous properties: it is very efficient, allowing its use for solving problems with many variables; it uses methods from linear algebra, which are readily solvable.

In January 1952 the first successful solution to a linear programming (LP) problem was found using a high-speed electronic computer on the National Bureau of Standards SEAC machine. Today, most LP's are solved via high-speed computers. Computer specific software, such as LINDO, EXCEL SOLVER, GAMS, have been developed to help in the solving and analysis of LP problems. We will use the power of LINDO to solve our linear programming problems in this chapter.

To provide a framework for our discussions, we offer the following basic model:

Maximize (or minimize) $f(X)$

Subject to

$$g_i(X) \begin{Bmatrix} \geq \\ = \\ \leq \end{Bmatrix} b_i \text{ for all } i.$$

Now lets' explain this notation. The various component of the vector X are called the decision variables of the model. These are the variables that can be controlled or manipulated. The function, $f(X)$, is called the objective function. By subject to, we connote that there are certain side conditions, resource requirement, or resource limitations that must be met. These conditions are called constraints. The constant b_i represents the level that the associated constraint $g(X_i)$ and is called the right-hand side in the model.

Linear programming is a method for solving linear problems, which occur very frequently in almost every modern industry. In fact, areas using linear programming are as diverse as defense, health, transportation, manufacturing, advertising, and telecommunications. The reason for this is that in most situations, the classic economic problem exists - you want to maximize output, but you are competing for limited resources. The 'Linear' in Linear Programming means that in the case of production, the quantity produced is proportional to the resources used and also the revenue generated. The coefficients are constants and no products of variables are allowed.

In order to use this technique the company must identify a number of constraints that will limit the production or transportation of their goods; these may include factors such as labor hours, energy, and raw materials. Each constraint must be quantified in terms of one unit of output, as the problem solving method relies on the constraints being used.

An optimization problem that satisfies the following five properties is said to be a linear programming problem.

- There is a unique objective function, $f(X)$.

- Whenever a decision variable, X, appears in either the objective function or a constraint function, it must appear with an exponent of 1, possibly multiplied by a constant.

- No terms contain products of decision variables.

- All coefficients of decision variables are constants.

- Decision variables are permitted to assume fractional as well as integer values.

Linear problems, by the nature of the many unknowns, are very hard to solve by human inspection, but methods have been developed to use the power of computers to do the hard work quickly. We will illustrate with two variables, graphically.

Supply Chain Management: A company owns railroad freight cars that can be sent all over the country. They need to work out what movements will be the most efficient in order to meet current customer needs and future needs on a probability basis, while minimizing time taken and costs incurred.

2. Formulating linear programming problems

A linear programming problem is a problem that requires an objective function to be maximized or minimized subject to resource constraints. The key to formulating a linear programming problem is recognizing the decision variables. The objective function and all constraints are written in terms of these decision variables.

The conditions for a mathematical model to be a linear program (LP) were:

- all variables continuous (i.e. can take fractional values)
- a single objective (minimize or maximize)
- the objective and constraints are linear i.e. any term is either a constant or a constant multiplied by an unknown.
- The decision variables must be non-negative

LP's are important - this is because:

- many practical problems can be formulated as LP's
- there exists an algorithm (called the *simplex* algorithm) that enables us to solve LP's numerically relatively easily.

We will return later to the simplex algorithm for solving LP's but for the moment we will concentrate upon formulating LP's.

Some of the major application areas to which LP can be applied are:

- Blending
- Production planning
- Oil refinery management
- Distribution
- Financial and economic planning
- Manpower planning
- Blast furnace burdening
- Farm planning

We consider below some specific examples of the types of problem that can be formulated as LP's. Note here that the key to formulating LP's is *practice*. However a useful hint is that common objectives for LP's are *minimize cost/maximize profit*.

Example 1 Manufacturing

Consider the following problem statement:

A company wants to can two new different drinks for the holiday season. It takes 2 hours to can one gross of Drink A, and it takes 1 hour to label the cans. It takes 3 hours to can one gross of Drink B, and it takes 4 hours to label the cans. The company makes $10 profit on one gross of Drink A and a $20 profit of one gross of Drink B. Given that we have 20 hours to devote to canning the drinks and 15 hours to devote to labeling cans per week, how many cans of each type drink should the company package to maximize profits?

Required Submission for Formulation solution:

Problem Identification: Maximize the profit of selling these new drinks.

Define variables:

X_1=the number of gross cans produced for Drink A per week

X_2= the number of gross cans produced for Drink B per week

Objective Function:

$Z=10X_1+20X_2$

      1. Canning with only 20 hours available per week

$$2X_1 + 3X_2 \le 20$$

      2. Labeling with only 15 hours available per week

$$X_1 + 4X_2 \le 15$$

      3. Non-negativity restrictions

            $X_1 \ge 0$ (non-negativity of the production items)

            $X_2 \ge 0$ (non-negativity of the production items)

The Complete FORMULATION:

MAXIMIZE $Z = 10X_1 + 20X_2$

subject to

$$2X_1 + 3X_2 \le 20$$

$$X_1 + 4X_2 \le 15$$

$$X_1 \ge 0$$

$$X_2 \ge 0$$

We will see in the next section how to solve these two-variable problems graphically.

Example 2 Financial planning

A bank makes four kinds of loans to its personal customers and these loans yield the following annual interest rates to the bank:

- First mortgage 14%

- Second mortgage 20%

- Home improvement 20%

- Personal overdraft 10%

The bank has a maximum foreseeable lending capability of \$250 million and is further constrained by the policies:

1. first mortgages must be at least 55% of all mortgages issued and at least 25% of all loans issued (in \$ terms)

2. second mortgages cannot exceed 25% of all loans issued (in \$ terms)

3.   to avoid public displeasure and the introduction of a new windfall tax the average interest rate on all loans must not exceed 15%.

Formulate the bank's loan problem as an LP so as to maximize interest income while satisfying the policy limitations.

Note here that these policy conditions, while potentially limiting the profit that the bank can make, also limit its exposure to risk in a particular area. It is a fundamental principle of risk reduction that risk is reduced by spreading money (appropriately) across different areas.

2.1. Financial planning formulation

Note here that as in *all* formulation exercises we are translating a verbal description of the problem into an *equivalent* mathematical description.

A useful tip when formulating LP's is to express the variables, constraints and objective in words before attempting to express them in mathematics.

2.2. Variables

Essentially we are interested in the amount (in dollars) the bank has loaned to customers in each of the four different areas (not in the actual number of such loans). Hence let

x_i= amount loaned in area i in a million of dollars (where i=1 corresponds to first mortgages, i=2 to second mortgages etc) and note that each $x_i >= 0$ (i=1,2,3,4).

Note here that it is conventional in LP's to have all variables >= 0. Any variable (X, say) which can be positive *or* negative can be written as X_1-X_2 (the difference of two new variables) where $X_1 >= 0$ and $X_2 >= 0$.

2.3. Constraints

a.   limit on amount lent

$x_1 + x_2 + x_3 + x_4 \leq 250$

b.   policy condition 1

$x_1 \geq 0.55(x_1 + x_2)$

i.e. first mortgages >= 0.55(total mortgage lending) and also

$x_1 \geq 0.25(x_1 + x_2 + x_3 + x_4)$

i.e. first mortgages $\geq$ 0.25(total loans)

c.   policy condition 2

$x_2 \leq 0.25(x_1 + x_2 + x_3 + x_4)$Inline formula

d. policy condition 3 - we know that the total annual interest is $0.14x_1 + 0.20x_2 + 0.20x_3 + 0.10x_4$ on total loans of $(x_1 + x_2 + x_3 + x_4)$. Hence the constraint relating to policy condition (3) is

$$0.14x_1 + 0.20x_2 + 0.20x_3 + 0.10x_4 \leq 0.15(x_1 + x_2 + x_3 + x_4) \tag{1}$$

2.4. Objective function

To maximize interest income (which is given above) i.e.

Maximize $Z = 0.14x_1 + 0.20x_2 + 0.20x_3 + 0.10x_4$

Example 3 Blending and Formulation

Consider the example of a manufacturer of animal feed who is producing feed mix for dairy cattle. In our simple example the feed mix contains two active ingredients. One kg of feed mix must contain a minimum quantity of each of four nutrients as below:

Nutrient	A	B	C	D
gram	90	50	20	2

The ingredients have the following nutrient values and cost

	A	B	C	D	Cost/kg
Ingredient 1 (gram/kg)	100	80	40	10	40
Ingredient 2 (gram/kg)	200	150	20	0	60

What should be the amounts of active ingredients in one kg of feed mix that minimizes cost?

2.5. Blending problem solution

Variables

In order to solve this problem it is best to think in terms of one kilogram of feed mix. That kilogram is made up of two parts - ingredient 1 and ingredient 2:

x_1= amount (kg) of ingredient 1 in one kg of feed mix

x_2= amount (kg) of ingredient 2 in one kg of feed mix

where $x_1 \geq 0, x_2 \geq 0$

Essentially these variables (x_1 and x_2) can be thought of as the recipe telling us how to make up one kilogram of feed mix.

Constraints

• nutrient constraints

$100x_1 + 200x_2 >= 90$ (nutrient A)

$80x_1 + 150x_2 >= 50$ (nutrient B)

$40x_1 + 20x_2 >= 20$ (nutrient C)

$10x_1 >= 2$ (nutrient D)

- balancing constraint (an *implicit* constraint due to the definition of the variables)

$x_1 + x_2 = 1$

Objective function

Presumably to minimize cost, i.e.

Minimize $Z = 40x_1 + 60x_2$

This gives us our complete LP model for the blending problem.

Example 4 Production planning problem

A company manufactures four variants of the same table and in the final part of the manufacturing process there are assembly, polishing and packing operations. For each variant the time required for these operations is shown below (in minutes) as is the profit per unit sold.

Variant 1	Assembly	Polish	Pack	Profit ($)
	2	3	2	1.50
2	4	2	3	2.50
3	3	3	2	3.00
4	7	4	5	4.50

Given the current state of the labor force the company estimate that, each year, they have 100,000 minutes of assembly time, 50,000 minutes of polishing time and 60,000 minutes of packing time available. How many of each variant should the company make per year and what is the associated profit?

Variables

Let:

x_i be the number of units of variant i ($i=1,2,3,4$) made per year, where $x_i \geq 0$ $i=1,2,3,4$

Constraints

Resources for the operations of assembly, polishing, and packing

$2x_1 + 4x_2 + 3x_3 + 7x_4 <= 100,000$ (assembly)

$3x_1 + 2x_2 + 3x_3 + 4x_4 < = 50,000$ (polishing)

$2x_1 + 3x_2 + 2x_3 + 5x_4 < = 60,000$ (packing)

Objective function

Maximize $Z = 1.5x_1 + 2.5x_2 + 3.0x_3 + 4.5x_4$

Example 5 Shipping

Consider planning the shipment of needed items from the warehouses where they are manufactured and stored to the distribution centers where they are needed as shown in the introduction. There are three warehouses at different cities: Detroit, Pittsburgh and Buffalo. They have 250, 130 and 235 tons of paper accordingly. There are four publishers in Boston, New York, Chicago and Indianapolis. They ordered 75, 230, 240 and 70 tons of paper to publish new books.

There are the following costs in dollars of transportation of one ton of paper:

From \ To	Boston (BS)	New York (NY)	Chicago (CH)	Indianapolis (IN)
Detroit (DT)	15	20	16	21
Pittsburgh (PT)	25	13	5	11
Buffalo (BF)	15	15	7	17

Management wants you to minimize the shipping costs while meeting demand.

We define x_{ij} to be the travel from city i (1 is Detroit, 2 is Pittsburg, 3 is Buffalo) to city j (1 is Boston, 2 is New York, 3 is Chicago, and 4 is Indianapolis).

Minimize $Z = 15x_{11} + 20x_{12} + 16x_{13} + 21x_{14} + 25x_{21} + 13x_{22} + 5x_{23} + 11x_{24} + 15x_{31} + 15x_{32} + 7x_{33} + 17x_{34}$

Subject to:

$x_{11} + x_{12} + x_{13} + x_{14} \leq 250$ (availability in Detroit)

$x_{21} + x_{22} + x_{23} + x_{24} \leq 130$ (availability in Pittsburg)

$x_{31} + x_{32} + x_{33} + x_{34} \leq 235$ (availability in Buffalo)

$x_{11} + x_{21} + x_{31} \geq 75$ (demand Boston)

$x_{12} + x_{22} + x_{32} \geq 230$ (demand New York)

$x_{13} + x_{23} + x_{334} \geq 240$ (demand Chicago)

$x_{14} + x_{24} + x_{34} \geq 70$ (demand Indianapolis)

$x_{ij} \geq 0$

3. LP geometry

Many applications in business and economics involve a process called optimization. In optimization problems, you are asked to find the minimum or the maximum result. This section illustrates the strategy in graphical simplex of linear programming. We will restrict ourselves in this graphical context to two-dimensions. Variables in the simplex method are restricted to positive variables (for example $x \geq 0$).

A two-dimensional linear programming problem consists of a linear objective function and a system of linear inequalities called constraints. The objective function gives the linear quantity that is to be maximized (or minimized). The constraints determine the *set of feasible solutions*.

Memory chips for CPUs

Let's start with a manufacturing example. Suppose a small business wants to know how many of two types of high-speed computer chips to manufacturer weekly to maximize their profits. First, we need to define our decision variables. Let,

x_1= number of high speed chip type A to produce weekly

x_2= number of high speed chip type B to produce week

The company reports a profit of \$140 for each type A chip and \$120 for each type B chip sold. The production line reports the following information:

	Chip A	Chip B	Quantity available
Assembly time (hours)	2	4	1400
Installation time (hours)	4	3	1500
Profit (per unit)	140	120	

The constraint information from the table becomes inequalities that are written mathematical as:

$2x_1 + 4x_2 \leq 1400$ (assembly time)

$4x_1 + 3x_2 \leq 1500$ (installation time)

$x_1 \geq 0, x_2 \geq 0$

The profit equation is:

$Profit = 140x_1 + 120x_2$

The feasible region

The constraints of a linear program, which include any bounds on the decision variables, essentially shape the region in the x-y plane that will be the domain for the objective function prior to any optimization being performed. Every inequality constraint that is part of the formulation divides the entire space defined by the decision variables into 2 parts: the portion of the space containing points that violate the constraint, and the portion of the space containing points that satisfy the constraint.

It is very easy to determine which portion will contribute to shaping the domain. We can simply substitute the value of some point in either *half-space* into the constraint. Any point will do, but the origin is particularly appealing. Since there's only one origin, if it satisfies the constraint, then the *half-space* containing the origin will contribute to the domain of the objective function.

When we do this for each of the constraints in the problem, the result is an area representing the intersection of all the *half-spaces* that satisfied the constraints individually. This intersection is the domain for the objective function for the optimization. Because it contains points that satisfy all the constraints simultaneously, these points are considered feasible to the problem. Naturally, the common name for this domain is the *feasible region*.

Consider our constraints:

$2x_1 + 4x_2 \leq 1400$ (assembly time)
$4x_1 + 3x_2 \leq 1500$ (installation time)
$x_1 \geq 0, x_2 \geq 0$

For our graphical work we use the constraints: $x_1 \geq 0$, $x_2 \geq 0$ to set the region. Here, we are strictly in the x_1-x_2 plane (the first quadrant).

Let's first take constraint #1 (assembly time) in the first quadrant: $2x_1 + 4x_2 \leq 1400$

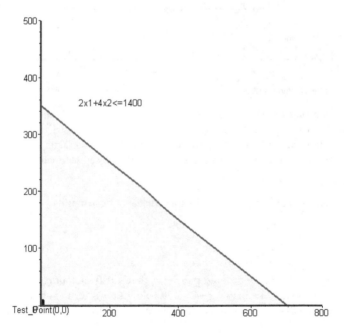

Figure 1. Shaded Inequality

We see the shaded region for constraint 1 that makes the inequality true. We repeat this process for all constraints to obtain Figure 2.

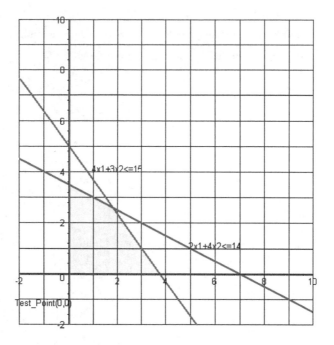

Figure 2. Plot of (1) the assembly hour's constraint and (2) the installation hour's constraint in the first quadrant

Figure 2 shows a plot of (1) the assembly hour's constraint and (2) the installation hour's constraint in the first quadrant. Along with the non-negativity restrictions on the decision variables, the intersection of the half-spaces defined by these constraints is the feasible region shown in yellow. This area represents the domain for the objective function optimization.

Finding the feasible region

Shade in the feasible region defined by the following set of constraints:

$x + 2y \leq 20$

$2x + y \leq 20$

$x \geq 0, y \geq 0$

The feasible region is the set of ordered pairs (x, y) that satisfy all four constraints simultaneously. They are points that lie below $x + 2y \leq 20$, below $2x + y \leq 20$, and above $y = 0$ and to the right of $x = 0$. We note that the non-negativity constraints, $x \geq 0, y \geq 0$, restrict the feasible region to the first quadrant. $\leq$

If the problem is well behaved, this should be a closed and bounded polyhedral shape, called a *polyhedron*, such as the one shown in yellow. It does not have to be so. Sometimes the orientation and location of the constraints fail to hold back the objective function in the direction of the optimization. When this happens, the problem is *unbounded*; the objective

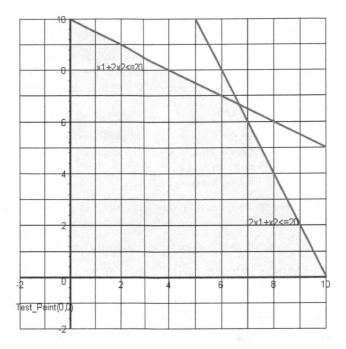

Figure 3. Shaded Feasible Region

function value goes off to positive or negative infinity. Can you draw a sketch of a situation in which this will happen?

Other times, the intersection of the half-spaces is an empty set. In this case, the problem is *infeasible*; there are no possible solutions that will satisfy the requirements of all the constraints simultaneously. Can you draw a sketch of a situation in which this will happen?

3.1. Solving a linear programming problem graphically

Recall that we have decision variables defined and an objection function that is to be maximized or minimized. Although all points inside the feasible region provide feasible solutions the solution, if one exists, occurs according to the Fundamental Theorem of Linear Programming:

If the optimal solution exists, then it occurs at a corner point of the feasible region.

Notice the various corners formed by the intersections of the constraints in example. These points are of great importance to us. There is a cool theorem (didn't know there were any of these, huh?) in linear optimization that states, "if an optimal solution exists, then an optimal corner point exists." The result of this is that any algorithm searching for the optimal solution to a linear program should have some mechanism of heading toward the corner point where the solution will occur. If the search procedure stays on the outside border of the feasible region while pursuing the optimal solution, it is called an *exterior point* method. If the search procedure cuts through the inside of the feasible region, it is called an *interior point* method.

Thus, in a linear programming problem, if there exists a solution, it must occur at a corner point of the set of feasible solutions (these are the vertices of the region). Note that in Figure 3 the corner points of the feasible region are the coordinates: $(0,0)$, $(0,10)$ $(10, 0)$, and $(20/3, 20/3)$.

How did we get the point $(20/3, 20/3)$?

This point is the intersection of the lines: $x + 2y = 20$ and $2x+y=20$. You have solved these problems before. In the second constraint, let $y = 20-2x$ and substitute $(20-2x)$ for y in the first equation, so $x + 2(20-2x) = 20$ and solve for x. We find $3x = 20$ or $x = 20/3$. Since $y = 20-2x$, we substitute $x = 20/3$ for x and solve for y. Now, $y = 20/3$.

Now, that we have all the possible solution coordinates for (x, y), we need to know which is the optimal solution. Here is how we determine that:

We evaluate the objective function at each point and choose the best solution.

Assume our objective function is to Maximize $Z = 2x + 2y$. We can set up a table of coordinates and corresponding Z-values as follows.

Coordinate of Corner Point	Z= 2x + 2y
(0,0)	Z= 0
(0,10)	$Z = (2)(0) + (2)(10) = 20$
(20/3, 20/3)	$Z=(2)(20/3)+(2)(20/3)= 80/3$ *
(10,0)	$Z = (2)(10) + (2)(0)= 20$
Best solution is (20/3,20/3)	$Z = 80/3 = 26.666$

Graphically, we see the result by plotting the objective function line, $Z = 2x + 2y$, with the feasible region. Determine the parallel direction for the line to maximize (in this case) Z. Move the line parallel until it crosses the last point in the feasible set. That point is the solution. The line that goes through the origin at a slope of -2/2 is called the ISO-Profit line. We have provided this Figure 4 below:

Here are the steps for solving a linear programming problem involving only two variables.

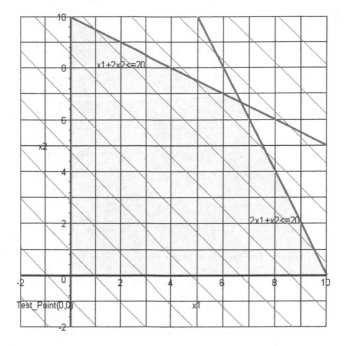

Figure 4. Iso-Profit Lines Added

1. Sketch the region corresponding to the system of constraints. The points satisfying all constraints make up the feasible solution.

2. Find all the corner points (or intersection points in the feasible region).

3. Test the objective function at each corner point and select the values of the variables that optimize the objective function. For bounded regions, both a maximum and a minimum will exist. For an unbounded region, if a solution exists, it will exist at a corner.

3.2. Minimization problem

Minimize $Z = 5x + 7y$

Subject to:

$2x + 3y \geq 6$
$3x - y \leq 15$
$-x + y \leq 4$
$2x + 5y \leq 27$
$x \geq 0$
$y \geq 0$

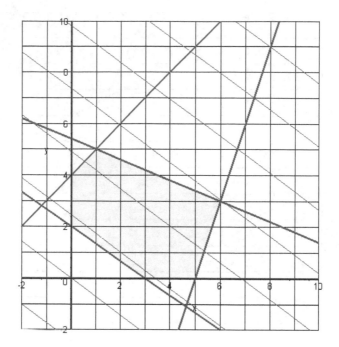

The corner points in Figure 5 are *(0,2), (0,4,) (1,5), (6,3), (5,0), and (3,0)*. See if you can find all these corner points.

If we evaluate $Z = 5x + 7y$ at each of these points, we find:

Corner Point	Z = 5x + 7y (MINIMIZE)
(0,2)	Z=14
(1,5)	Z= 40
(6,3)	Z= 51
(5,0)	Z=25
(3,0)	Z=15
(0,4)	Z=28

The minimum value occurs at (0, 2) with a Z value of 14. Notice in our graph that the blue ISO-Profit line will last cross the point (0,2) as it move out of the feasible region in the direction that Minimizes Z.

3.3. Unbounded case

Let's examine the concept of an unbounded feasible region. Look at the constraints:

$x + 2y \geq 4$

$3x + y \geq 7$

$x \geq 0$ and $y \geq 0$

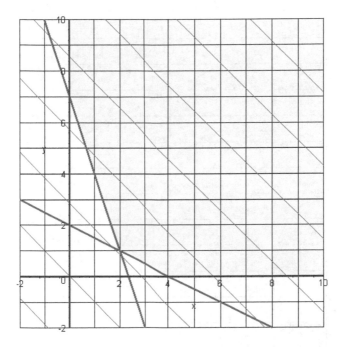

Note that the corner points are $(0, 7)$, $(2, 1)$ and $(4, 0)$ and the region is unbounded. If our solution is to Minimize $Z = x + y$ then our solution is: $(2, 1)$ with $Z = 3$. Determine why there is no solution to the LP to Maximize $Z = x + y$.

4. Graphical sensitivity analysis

One of the most important topics in linear programming is sensitivity analysis. In this section we illustrate the concept of sensitivity analysis through a graphical example. Sensitivity analysis is concerned with how changes in the parameters (coefficient and right-hand-side values) affect the LP's optimal solution. Very often, we can ascertain whether the optimal solution variables remain the same (perhaps with different solution values) or whether the variables will change.

Reconsider the following example:

Maximize 2x+ 2y

Subject to:

x+ 2y ≤ 20
2x+y ≤ 20
x ≥ 0, y ≥ 0

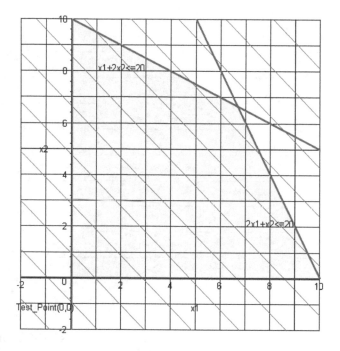

The corners of the feasible region with the respective objective function values are shown in the following table:

Point	Coordinates	Z-Value
A	(0,0)	Z=0
B	(10,0)	Z=20
C	(0,10)	Z=20
D	(20/3,20/3)	Z= 80/3=26.66666 Optimal

The optimal solution is currently x= 20/3, y = 20/3 and Z = 80/3.

4.1. Graphical analysis of the effect of an objective function coefficient

Recall our objective function:

$Z= 2x+ 2y$

The slope of this objective function line is –1. The solution is currently at Point D, the coordinates (20/3, 20/3). Recall that the optimal solution, if it exists, must lie at the corner point of the feasible region. The next closest points to D are B and C. B lies on constraint (2) and C lies on constraints (1). The slope of constraint (1) is –1/2 and the slope of constraint (2) is –2.

You can see that if we change the coefficients, let say, A $x + 2 y$, its slope is –A/2. Currently, A is 2 with a slope of –1. The bounds for the slope to retain Point D as a solution is between –2 ≤ slope ≤ –1/2 (Note at equality we will have alternate optimal solutions). We can easily solve for the values of A that keep the slope within its bounds. 1≤A≤4. Since A is currently 2, it can decrease by 1 unit or increase by 2 units and still keep Point D optimal.

Let's try 2 $x +$ by with its slope –2/b. Currently, B is 2 with a slope of –1. The bounds for the slope to retain Point D as a solution is between –2 ≤ slope ≤ –1/2 (Note at equality we will have alternate optimal solutions). We can easily solve for the values of B that keep the slope within its bounds. 1≤B≤4. Since B is currently 2, it can decrease by 1 unit or increase by 2 units and still keep Point D optimal.

4.2. Graphical analysis of the effect of a change in a right-hand side coefficient

The right-hand side value for each constraint controls the y-intercept of the problem. The slopes will remain the same. Changing the right-hand side yields a parallel line to the original constraint. Point D is the intersection of constraints (1) and (2). Points B and C are the intersections of constraints (1) and Constraint (2) each with non-negativity.

Let's consider constraint (1), $x+ 2y \le B$

The value of B is currently 20. If it is reduced then the line moves down until it intersects point (10, 0). Using (10, 0) in the equation yields a B value of 10. If we increase B, then we move up to an original infeasible point (0, 20), which will now become feasible. Using (0, 20) in the equation yields a B value of 40.

Thus, 10≤B≤40. This is a decrease of 10 units and an increase of 20 units.

Let's consider constraint (2), $2x + y \le C$

The value of C is currently 20. If it is reduced then the line moves down until it intersects point (0, 10). Using (0, 10) in the equation yields a C value of 10. If we increase C, then we move up to an original infeasible point (20, 0), which will now become feasible. Using (20, 0) in the equation yields a C value of 40.

Thus, 10≤C≤40. This is a decrease of 10 units and an increase of 20 units.

Maximize $Z= 20x_1+30 x_2$

Subject to:

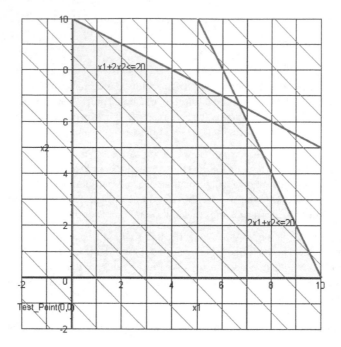

$2x_1 + 4x_3 \leq 1400$ (assembly time)

$4x_1 + 3x_3 \leq 1500$ (installation time)

The solution is (180,260), Z =11400.

Let's see the effects of changing the coefficients of the objective function since management has some leeway with the revenues and costs

$Z = 20x_1 + 30x_2$

The slope of this objective function line is –2/3. The solution is currently at Point D, the coordinates (180,260). Recall that the optimal solution, if it exists, must lie at the corner point of the feasible region. The next closest points to D are B and C. B lies on constraint (2) and C lies on constraints (1). The slope of constraint (1) is –1/2 and the slope of constraint (2) is –4/3.

You can see that if we change the coefficients, let say, A x_1 + 30 x_2, its slope is –A/30. Currently, A is 20 with a slope of –2/3. The bounds for the slope to retain Point D has a solution between –4/3 ≤ slope≤ –1/2. We can easily solve for the values of A that keep the slope within its bounds. 15≤A≤40. Since A is currently 20, it can decrease by 5 units or increase by 20 units and still keep Point D optimal.

Now, let's try changing the coefficient of the other variable: 20 x + B y with its slope –20/B. Currently, B is 30 with a slope of –2/3. The bounds for the slope to retain Point D as a solution is between –4/3 ≤ slope≤ –1/2 We can easily solve for the values of B that keep the slope within

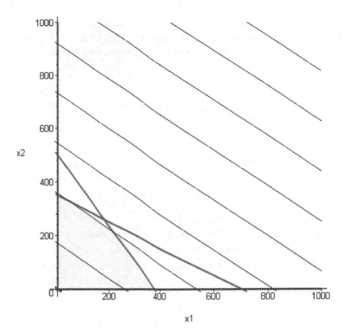

its bounds. 15≤B≤40. Since B is currently 30, it can decrease by 15 units or increase by 10 units and still keep Point D optimal.

5. The simplex method

In the previous sections we discussed formulating linear programming problems, solving two-dimensional linear programming problems by graphical methods, and graphical sensitivity analysis. The graphical method illustrates some key concepts, but is only practical for problems with two variables. As you see linear programming problems often have more than two variables. With problems with more than two variables, an algebraic method may be used. This method is called the Simplex Method. The *Simplex Method*, developed by George Dantzig in 1947 incorporates both *optimality* and *feasibility* tests to find the optimal solution(s) to a linear program (if one exists).

An *optimality test* shows whether or not an intersection point corresponds to a value of the objective function better than the best value found so far.

A *feasibility test* determines whether the proposed intersection point is feasible. It does not violate any of the constraints.

The simplex method starts with the selection of a corner point (usually the origin if it is a feasible point) and then, in a systematic method, moves to adjacent corner points of the feasible region until the optimal solution is found or it can be shown that no solution exists.

5.1. Steps of the simplex method

1. *Tableau Format*: Place the linear program in Tableau Format, as explained below.

Maximize $Z = 25x_1 + 30x_2$

Subject to:

$20x_1 + 30x_2 \leq 6905$

$x_1 + 4x_2 \leq 120$

$x_1, x_2, \geq 0$

To begin the simplex method, we start by converting the inequality constraints (of the form <) to equality constraints. This is accomplished by adding a unique, non-negative variable, called a slack variable, to each constraint. For example the inequality constraint $20x_1 + 30x_2 \leq 690$ is converted to an equality constraint by adding the slack variable S_1 to obtain:

$20x_1 + 30x_2 + S_1 = 690$, where $S_1 \geq 0$.

The inequality $20x_1 + 30x_2 \leq 690$ states that the sum $20x_1 + 30x_2$ is less than or equal to 690. The slack variable "takes up the slack" between the values used for x_1 and x_2 and the value 690. For example, if $x_1 = x_2 = 0$, the $S_1 = 690$. If $x_1 = 24$, $x_2 = 0$ then $20(24) + 30(0) + S_1 = 690$, so $S_1 = 210$.

A unique slack variable must be added to each inequality constraint.

Maximize $Z = 25x_1 + 30x_2$

Subject to:

$20x_1 + 30x_2 + S_1 = 690$

$5x_1 + 4x_2 + S_2 = 120$

$x_1 \geq 0, x_2 \geq 0, S_1 \geq 0, S_2 \geq 0$

Adding slack variables makes the constraint set a system of linear equations. We write these with all variables on the left side of the equation and all constants on the right hand side.

We will even rewrite the objective function by moving all variables to the left-hand side.

Maximize $Z = 25x_1 + 30x_2$ is written as

$Z - 25x_1 - 30x_2 = 0$

Now, these can be written in the following form:

$Z - 25x_1 - 30x_2 = 0$

$20x_1 + 30x_2 + S_1 = 690$

$5x_1 + 4x_2 + S_2 = 120$

or more simply in a matrix. This matrix is called the simplex tableau.

Z	x_1	x_2	S_1	S_2		RHS
1	-25	-30	0	0	=	0
0	20	30	1	0	=	690
0	5	4	0	1	=	120

2. *Initial Extreme Point*: The Simplex Method begins with a known extreme point, usually the origin (0, 0) for many of our examples. The requirement for a basic feasible solution gives rises to special Simplex methods such as Big M and Two-Phase Simplex, which can be studied in a linear programming course.

The Tableau previously shown contains the corner point (0, 0) is our initial solution.

Z	x_1	x_2	S_1	S_2		RHS
1	-25	-30	0	0	=	0
0	20	30	1	0	=	690
0	5	4	0	1	=	120

We read this solution as follows:

$x_1 = 0$

$x_2 = 0$

$S_1 = 690$

$S_2 = 120$

$Z = 0$

Let's continue to define a few of these variables further. We have 5 variables $\{Z, x_1, x_2, S_1, S_2\}$ and 3 equations. We can have at most 3 solutions. Z will always be a solution by convention of our tableau. We have two non-zero variables among $\{x_1, x_2, S_1, S_2\}$. These non-zero variables are called the *basic variables*. The remaining variables are called the *non-basic variables*. The corresponding solutions are called the *basic feasible solutions* (FBS) and correspond to corner points. The complete step of the simplex method produces a solution that corresponds to a corner point of the feasible region. These solutions are read directly from the tableau matrix.

We also note the basic variables are variables that have a column consisting of one 1 and the rest zeros in their column. We will add a column to label these as shown below:

	Basic Variable		Basic variable	Basic variable			
	Z	x_1	x_2	S_1	S_2		RHS
Z	1	-25	-30	0	0	=	0

Basic Variable				Basic variable	Basic variable		
S_1	0	20	30	1	0	=	690
S_2	0	5	4	0	1	=	120

3. *Optimality Test*: We need to determine if an adjacent intersection point improves the value of the objective function. If not, the current extreme point is optimal. If an improvement is possible, the optimality test determines which variable currently in the independent set (having value zero) should *enter* the dependent set as a basic variable and become nonzero. For our maximization problem, we look at the Z-Row (The row marked by the basic variable Z). If any coefficients in that row are negative then we select the variable whose coefficient is the most negative as the entering variable.

Basic Variable				Basic variable	Basic variable		
	Z	x_1	x_2	S_1	S_2		RHS
Z	1	-25	-30	0	0	=	0
S_1	0	20	30	1	0	=	690
S_2	0	5	4	0	1	=	120

In the Z-Row the coefficients are:

	Z	x_1	x_2	S_1	S_2
Z	1	-25	-30	0	0

The variable with the most negative coefficient is x_2 with value –30. Thus, x_2 wants to become a basic variable. We can only have three basic variables in this example (because we have three equations) so one of the current basic variables $\{S_1, S_2\}$ must be replaced by x_2. Let's proceed to see how we determine which variable exists being a basic variable.

4. *Feasibility Test*: To find a new intersection point, one of the variables in the basic variable set must *exit* to allow the entering variable from Step 3 to become basic. The feasibility test determines which current dependent variable to choose for exiting, ensuring we stay inside the feasible region. We will use the minimum positive ratio test as our feasibility test. The Minimum Positive Ratio test is the MIN($RHS_j/a_j > 0$). Make a quotient of the $\dfrac{rh\,s_j}{a_j}$.

	Z	x_1	Most negative coefficient (-30) x_2	S_1	S_2		RHS	Ratio Test Quotient
Z	1	-25	-30	0	0	=	0	
S_1	0	20	30	1	0	=	690	690/30=23
S_2	0	5	4	0	1	=	120	120/4=30

Note that we will always disregard all quotients with either 0 or negative values in the denominator. In our example we compare {23, 30} and select the smallest non-negative value. This gives the location of the matrix pivot that we will perform.

	Z	x_1	Most negative coefficient (-30) x_2	S_1	S_2		RHS	Ratio Test Quotient
Z	1	-25	-30	0	0	=	0	
S_1	0	20	30 Pivot	1	0	=	690	690/30=23
S_2	0	5	4	0	1	=	120	120/4=30

5. *Pivot:* We can form a new equivalent system by using row operations to change the pivot element to a 1 and all other numbers in the pivot column to zero. We do the row operations by adding a suitable multiple of the pivot row to a multiple of each row in the tableau, thus eliminating the new basic variable. Then set the new non-basic variables to zero in the new system to find the values of the new basic variables, thereby determining an intersection point.

	Z	x_1	Pivot Column x_2	S_1	S_2		RHS	
Z	1	-25	-30	0	0	=	0	
S_1	0/30	20/30	30/30	1/30	0/30	=	690/30	Pivot Row
S_2	0	5	4	0	1	=	120	

Make the entry in the intersection of the pivot column and pivot row equal to 1.

	Z	x_1	Pivot Column x_2	S_1	S_2		RHS	
Z	1	-25	-30	0	0	=	0	
S_1	0	2/3	1	1/30	0/	=	23	Pivot Row
S_2	0	5	4	0	1	=	120	

Using row operations make all other entries in the pivot column equal to 0.

	Z	x_1	Pivot Column x_2	S_1	S_2		RHS	
Z	1	-5	0	1	0	=	690	$30R_2+R_1 àR_1$
x_2	0	2/3	1	1/30	0	=	23	
S_2	0	7/3	0	-4/30	1	=	28	$-4R_2+R_3 àR_3$

Let's interpret our current basic feasible solution.

Basic Variables:

x_2=23

S_2=32

Z= 690

Non-Basic Variables $x_1 = 0, S_1 =0$

6. Repeat Steps 3-5 until an optimal extreme point is found.

We note that x_1 has a coefficient of –5 in the Z-Row therefore, we are not optimal.

Step 3.

	Z	x_1	Negative Coefficient x_2	S_1	S_2		RHS
Z	1	-5	0	1	0	=	690
x_2	0	2/3	1	1/30	0	=	23
S_2	0	7/3	0	-4/30	1	=	28

Step 4.

	Z	x_1	x_2	S_1	S_2		RHS	Quotient
		Pivot Column						Ratio Test
Z	1	-5	0	1	0	=	690	
x_2	0	2/3	1	1/30	0	=	23	69/2=34.5
S_2	0	7/3	0	-4/30	1	=	28	84/7=12* Min

The minimum non-negative quotient is 12. This indicates that to remain in the feasible region that x_1 enters as a basic variable and S_2 leaves being a basic variable.

	Z	x_1	x_2	S_1	S_2		RHS	Quotient
		Pivot Column						Ratio Test
Z	1	-5	0	1	0	=	690	
x_2	0	2/3	1	1/30	0	=	23	
S_2	0	7/3	0	-4/30	1	=	28	Pivot Row

Step 5. We make the highlighted position a 1 and all other column entries 0 for the column of x_1. We divide the entire S_2 row by 7/3.

	Z	x_1	x_2	S_1	S_2		RHS
Z	1	-5	0	1	0	=	690
x_2	0	2/3	1	1/30	0	=	23
S_2	0	1	0	-12/210	3/7	=	12

	Z	x_1	x_2	S_1	S_2		RHS	
Z	1	0	0	5/7	15/7	=	750	$5R_3+R_1 àR_1$
x_2	0	0	1	1/14	-2/7	=	15	$-2/3R_3+R_2 àR_2$
x_1	0	1	0	-2/35	3/7	=	12	

The current solution is read as follows:

Basic Variables

$x_2 = 15$

$x_1 = 12$

$Z = 750$

Non-basic variables

$S_1 = S_2 = 0$

There are no negative coefficients in the Z-Row, so we are optimal.

	Z	x_1	x_2	S_1	S_2
Z	1	0	0	5/7	15/7

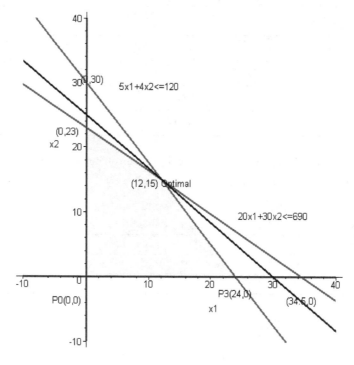

Figure 5. The set of points satisfying the constraint of this linear programming problem (the convex set as a shaded region)

Each solution found in the tableau corresponds to a corner point. We went from corner $(0, 0)$ to corner $(0, 23)$ to corner $(12, 15)$.

Maximize $Z = 3x_1 + x_2$

Subject to:

$2x_1 + x_2 \leq 6$

$x_1 + 3x_2 \leq 9$

$x_1, x_2 \geq 0$

The Tableau Format with slack variables y_1, y_2:

Basic Var.	Z	x_1	x_2	y_1	y_2	RHS
Z	1	-3	-1	0	0	0
y_1	0	2	1	1	0	6
y_2	0	1	3	0	1	9

Basic variable $\{Z, y_1, y_2\}$

Non-basic Variable $\{x1, x2\}$

Extreme Point $(0, 0)$ ---Corresponding to the values of (x_1, x_2)

Value of objective function: $Z=0$

Optimality Test: The entering variable is x_1 (corresponding to -3 in the Z- row.)

Feasibility Test: Compute the ratios of the "RHS" divided by the column labeled x_1 to determine the minimum positive ratio.

Basic Var.	Z	x_1	x_2	y_1	y_2	RHS	Quotient
Z	1	-3	-1	0	0	0	
y_1	0	2	1	1	0	6	6/2=3
y_2	0	1	3	0	1	9	9/1=9

Choose y_1 to leave since it corresponds to the minimum positive ratio test value of 3.

Pivot: Divide the row containing the exiting variable (the first row in this case) by the coefficient of the entering variable in that row (the coefficient of x_1 in this case), giving a coefficient of 1 for the entering variable in this row. Then eliminate the entering variable x_1 from the remaining rows (which do not contain the exiting variable y_1 and have a zero coefficient for it). The results are summarized in the next tableau.

Basic Var.	Z	x	x_2	y_1	y_2	RHS
Z	1	0	0.5000	1.5	0	9
x_1	0	1	0.5000	.5	0	3
y_2	0	0	2.5000	-.5	1	6

Basic variable $\{Z, x_1, y2\}$

Non-basic Variable $\{y_1, x_2\}$

Extreme Point (3, 0) ---Corresponding to the values of (x_1, x_2)

Value of objective function: $Z=9$

The pivot determines that the dependent variables have the values $x_1=3$, $y_2=6$ and $Z=9$.

Optimality Test: There are no negative coefficients in the Z- row. Thus $x_1=3$ (a basic variable) and $x_2=0$ (a non-basic variable) is an extreme point giving the optimal objective function value $Z=9$.

We read the solution as $Z=9$, $x_1=3$, and $s_2=6$.

Remarks: We have assumed that the origin is a feasible extreme point. If it is not, then an extreme point must be found before the Simplex Method, as presented, can be used. We have also assumed that the linear program is not ``degenerate" in the sense that no more than two constraints intersect at the same point. These and other topics are studied in more advanced optimization courses.

6. Linear programming with technology

Technology is critical to solving, analyzing, and performing sensitivity analysis on linear programming problems. Technology provides a suite of powerful, robust routines for solving optimization problems, including linear programs (LPs). Technology, that we briefly discuss, includes Excel, LINDO, and LINGO as these appear to be used often in engineering. We tested all these software packages and found them useful.

We present our previous example solved via each technology.

Maximize $Z = 25x_1+30x_2$

Subject to:

$20x_1+30x_2 \leq 6905$

$x_1+4x_2 \leq 120$

$x1, x2, \geq 0$

EXCEL

	A	B	C	D	E	F
1	LP in EXCEL					
2						
3						
4	Decision	Variables			Objective Function	
5		Initial/Final Values			=28*B6+30*B7	
6	x1	0				
7	x2	0				
8						
9						

ont   Alignment

=2*B6+3*B7

	A	B	C	D	E	F
1	LP in EXCEL					
2						
3						
4	Decision	Variables			Objective Function	
5		Initial/Final Values			0	
6	x1	0				
7	x2	0				
8						
9						
10	Constraints			Used	RHS	
11				0	100	
12				0	120	
13				0	90	
14						
15						

Solver

Constraints into solver

	A	B	C	D	E	F
1	LP in EXCEL					
2						
3						
4	Decision	Variables			Objective Function	
5		Initial/Final Values			972	
6	x1	9				
7	x2	24				
8						
9						
10	Constraints			Used	RHS	
11				90	100	
12				120	120	
13				90	90	
14						

Reports

Answer Report

	B	C	D	E	F	G	H
1	Microsoft Excel 14.0 Answer Report						
2	Worksheet: [Book3]Sheet1						
3	Report Created: 8/28/2012 1:37:02 PM						
4	Result: Solver has converged to the current solution. All Constraints are satisfied.						
5	Solver Engine						
6	Engine: Simplex LP						
7	Solution Time: 0.015 Seconds.						
8	Iterations: 2 Subproblems: 0						
9	Solver Options						
10	Max Time Unlimited, Iterations Unlimited, Precision 0.000001, Use Automatic Scaling						
11	Max Subproblems Unlimited, Max Integer Sols Unlimited, Integer Tolerance 1%, Assume NonNegative						
12							
13							
14	Objective Cell (Max)						
15	Cell	Name	Original Value	Final Value			
16	$E$5	Initial/Final Values Objective Function	0	972			
17							
18							
19	Variable Cells						
20	Cell	Name	Original Value	Final Value	Integer		
21	$B$6	x1 Initial/Final Values	0	9	Contin		
22	$B$7	x2 Initial/Final Values	0	24	Contin		
23							
24							
25	Constraints						
26	Cell	Name	Cell Value	Formula	Status	Slack	
27	$D$11	Used	90	$D$11<=$E$11	Not Binding	10	
28	$D$12	Used	120	$D$12<=$E$12	Binding	0	
29	$D$13	Used	90	$D$13<=$E$13	Binding	0	

Sensitivity Report

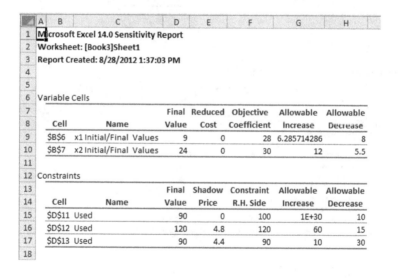

A	B	C	D	E	F	G	H
1	Microsoft Excel 14.0 Sensitivity Report						
2	Worksheet: [Book3]Sheet1						
3	Report Created: 8/28/2012 1:37:03 PM						
4							
5							
6	Variable Cells						
7			Final	Reduced	Objective	Allowable	Allowable
8	Cell	Name	Value	Cost	Coefficient	Increase	Decrease
9	$B$6	x1 Initial/Final Values	9	0	28	6.285714286	8
10	$B$7	x2 Initial/Final Values	24	0	30	12	5.5
11							
12	Constraints						
13			Final	Shadow	Constraint	Allowable	Allowable
14	Cell	Name	Value	Price	R.H. Side	Increase	Decrease
15	$D$11	Used	90	0	100	1E+30	10
16	$D$12	Used	120	4.8	120	60	15
17	$D$13	Used	90	4.4	90	10	30
18							

We find our solution is $x_1=9$, $x_2=24$, $P=\$972$. From the standpoint of sensitivity analysis Excel is satisfactory in that it provides shadow prices.

Limitation: No tableaus are provided making it difficult to find alternate solutions.

LINDO

```
MAX    25 X1 + 30 X2
 SUBJECT TO
    2)  20 X1 + 30 X2 <=  690
    3)  5 X1 + 4 X2 <=  120
END


THE TABLEAU
         ROW  (BASIS)      X1      X2  SLK   2  SLK   3
          1 ART     -25.000  -30.000   0.000   0.000    0.000
          2 SLK   2  20.000   30.000   1.000   0.000  690.000
          3 SLK   3   5.000    4.000   0.000   1.000  120.000
       ART     ART  -25.000  -30.000   0.000   0.000    0.000


LP OPTIMUM FOUND AT STEP    2
    OBJECTIVE FUNCTION VALUE
    1)    750.0000


VARIABLE      VALUE      REDUCED COST
    X1     12.000000     0.000000
    X2     15.000000     0.000000


    ROW  SLACK OR SURPLUS    DUAL PRICES
    2)     0.000000      0.714286
    3)     0.000000      2.142857


NO. ITERATIONS=     2


RANGES IN WHICH THE BASIS IS UNCHANGED:
         OBJ COEFFICIENT RANGES


VARIABLE       CURRENT      ALLOWABLE       ALLOWABLE


            COEF      INCREASE     DECREASE
    X1    25.000000   12.500000     5.000000
    X2    30.000000    7.500000    10.000000


              RIGHTHAND SIDE RANGES


    ROW      CURRENT      ALLOWABLE       ALLOWABLE


            RHS      INCREASE     DECREASE
    2   690.000000  210.000000  209.999985
    3   120.000000   52.499996   28.000000


THE TABLEAU


    ROW  (BASIS)      X1      X2  SLK   2  SLK   3
     1 ART      0.000    0.000   0.714   2.143  750.000
     2    X2    0.000    1.000   0.071  -0.286   15.000
     3    X1    1.000    0.000  -0.057   0.429   12.000
```

LINGO

MODEL:

MAX = 25 * x1 + 30 * x2;

    20 * x1 + 30 * x2 <= 690;
    5 * x1 + 4 * x2 <= 120;
    x1>=0;
    x2>=0;
END

Variable	Value	Reduced Cost
X1	12.00000	0.0000000
X2	15.00000	0.0000000

Row	Slack or Surplus	Dual Price
1	750.0000	1.000000
2	0.0000000	0.7142857
3	0.0000000	2.142857
4	12.00000	0.0000000
5	15.00000	0.0000000

7. Case study

In our case study we present linear programming for supply chain design. We consider producing a new mixture of gasoline. We desire to minimize the total cost of manufacturing and distributing the new mixture. There is a supply chain involved with a product that must be modeled. The product is made up of components that are produced separately.

Crude Oil type	Compound A (%)	Compound B (%)	Compound C (%)	Cost/Barrel	Barrel Avail (000 of barrels)
X10	35	25	35	$26	15000
X20	50	30	15	$32	32000
X30	60	20	15	$55	24000

Demand information is as follows:

Gasoline	Compound A (%)	Compound B (%)	Compound C (%)	Expected Demand (000 of barrels)
Premium	≥ 55	≤23		14000
Super		≥25	≤35	22000
Regular	≥40		≤25	25000

Let i = crude type 1, 2, 3 (X10,X20,X30 respectively)

Let j = gasoline type 1, 2, 3 (Premium, Super, Regular respectively)

We define the following decision variables:

Gij= amount of crude i used to produce gasoline j

For example,

G_{11}= amount of crudeX10used to produce Premium gasoline.

G_{12}= amount of crude typeX20used to produce Premium gasoline

G_{13}= amount of crude typeX30used to produce Premium gasoline

G_{12}= amount of crude typeX10used to produce Super gasoline

G_{22}= amount of crude typeX20used to produce Super gasoline

G_{32}= amount of crude typeX30used to produce Super gasoline

G_{13}= amount of crude typeX10used to produce Regular gasoline

G_{23}= amount of crude typeX20used to produce Regular gasoline

G_{33}= amount of crude typeX30used to produce Regular gasoline

LP formulation

Minimize Cost= \$86 (G11+G21+G31)+\$92(G12+G22+G32)+\$95(G13+G23+G33)

Subject to: Demand

G11+G21+G31>14000 (Premium)
G12+G22+G32>22000 (Super)
G13+G23+G33>25000 (Regular)

Availability of products

G11+G12+G13 < 15000 (crude 1)
G21+G22+G23 < 32000 (crude 2)
G31+G32+G33 < 24000 (crude 3)

Product mix in mixture format

$(0.35\,G11 + 0.50\,G21 + 0.60\,G31)/(G11 + G21 + G31) \geq 0.55\ (X\,10\ in\ Premium)$
$(0.25\,G11 + 0.30\,G21 + 0.20\,G31)/(G11 + G21 + G31) \leq 0.23\ (X\,20\ in\ Premium)$
$(0.35\,G13 + 0.15\,G23 + 0.15\,G33)/(G13 + G23 + G33) \geq 0.25\ (X\,20\ in\ Regular)$
$(0.35\,G13 + 0.15\,G23 + 0.15\,G33)/(G13 + G23 + G33) \leq 0.35\ (X\,30\ in\ Regular)$
$(0.35G12 + 0.50G22 + 0.60\,G23)/(G12 + G22 + G32) \leq 0.40\ (Compound\ X\,10\ in\ Super)$
$(0.35G12 + 0.15G22 + 0.15\,G32)/(G12 + G22 + G32) \leq 0.25\ (Compound\ X\,30\ in\ Super)$

The solution was found using LINDO and we noticed an alternate optimal solution. Two solutions are found yielding a minimum cost of \$1,904,000.

Decision variable	Z=$1,940,000	Z=$1,940,000
G_{11}	0	1,400
G_{12}	0	3,500
G_{13}	14,000	9,100
G_{21}	15,000	1,100
G_{22}	7,000	20,900
G_{23}	0	0
G_{31}	0	12,500
G_{32}	25,000	7,500
G_{33}	0	4,900

Depending on whether we want to additionally minimize delivery (across different locations) or maximize sharing by having more distribution point involved then we have choices.

We present one of the solutions below with LINDO.

LP OPTIMUM FOUND AT STEP 7		
OBJECTIVE FUNCTION VALUE		
1)	1904000.	
VARIABLE	VALUE	REDUCED COST
P1	0.000000	0.000000
R1	15000.000000	0.000000
E1	0.000000	0.000000
P2	0.000000	0.000000
R2	7000.000000	0.000000
E2	25000.000000	0.000000
P3	14000.000000	0.000000
R3	0.000000	0.000000
E3	0.000000	0.000000

ROW	SLACK OR SURPLUS	DUAL PRICES
2)	0.000000	9.000000
3)	0.000000	3.000000
4)	10000.000000	0.000000
5)	0.000000	-35.000000
6)	0.000000	-35.000000
7)	0.000000	-35.000000
8)	700.000000	0.000000
9)	3500.000000	0.000000
10)	1400.000000	0.000000
11)	2500.000000	0.000000
12)	2500.000000	0.000000
13)	420.000000	0.000000

NO. ITERATIONS= 7

RANGES IN WHICH THE BASIS IS UNCHANGED:

OBJ COEFFICIENT RANGES

VARIABLE	CURRENT COEF	ALLOWABLE INCREASE	ALLOWABLE DECREASE
P1	26.000000	INFINITY	0.000000
R1	26.000000	0.000000	INFINITY
E1	26.000000	INFINITY	0.000000
P2	32.000000	0.000000	0.000000
R2	32.000000	0.000000	0.000000
E2	32.000000	0.000000	35.000000
P3	35.000000	0.000000	3.000000
R3	35.000000	INFINITY	0.000000
E3	35.000000	INFINITY	0.000000

| | RIGHTHAND SIDE RANGES | | |
ROW	CURRENT RHS	ALLOWABLE INCREASE	ALLOWABLE DECREASE
2	15000.000000	4200.000000	0.000000
3	32000.000000	4200.000000	0.000000
4	24000.000000	INFINITY	10000.000000
5	14000.000000	10000.000000	14000.000000
6	22000.000000	0.000000	4200.000000
7	25000.000000	0.000000	4200.000000
8	0.000000	700.000000	INFINITY
9	0.000000	3500.000000	INFINITY
10	0.000000	INFINITY	1400.000000
11	0.000000	2500.000000	INFINITY
12	0.000000	INFINITY	2500.000000
13	0.000000	INFINITY	420.000000

THE TABLEAU

ROW	(BASIS)	P1	R1	E1	P2	R2	E2
1	ART	0.000	0.000	0.000	0.000	0.000	0.000
2	R1	1.000	1.000	1.000	0.000	0.000	0.000
3	P2	1.000	0.000	0.000	1.000	0.000	0.000
4	SLK	4	0.000	0.000	0.000	0.000	0.000
5	P3	0.000	0.000	0.000	0.000	0.000	0.000
6	R2	-1.000	0.000	-1.000	0.000	1.000	0.000
7	E2	0.000	0.000	1.000	0.000	0.000	1.000
8	SLK	8	0.150	0.000	0.000	0.000	0.000
9	SLK	9	-0.500	0.000	-0.500	0.000	0.000
10	SLK	10	-0.200	0.000	-0.200	0.000	0.000
11	SLK	11	0.000	0.000	0.150	0.000	0.000
12	SLK	12	0.000	0.000	0.200	0.000	0.000
13	SLK	13	-0.050	0.000	0.000	0.000	0.000

ROW	P3	R3	E3	SLK 2	SLK 3	SLK 4	SLK 5
1	0.000	0.000	0.000	9.000	3.000	0.000	35.000
2	0.000	0.000	0.000	1.000	0.000	0.000	0.000
3	0.000	-1.000	-1.000	1.000	1.000	0.000	0.000
4	0.000	0.000	0.000	1.000	1.000	1.000	1.000
5	1.000	1.000	1.000	-1.000	-1.000	0.000	-1.000
6	0.000	1.000	0.000	-1.000	0.000	0.000	0.000
7	0.000	0.000	1.000	0.000	0.000	0.000	0.000
8	0.000	0.100	0.100	-0.100	-0.100	0.000	-0.050
9	0.000	1.000	0.000	-0.500	0.000	0.000	0.000
10	0.000	0.000	0.000	-0.200	0.000	0.000	0.000
11	0.000	0.000	-0.100	0.000	0.000	0.000	0.000
12	0.000	0.000	0.000	0.000	0.000	0.000	0.000
13	0.000	0.100	0.100	-0.100	-0.100	0.000	-0.030

ROW	SLK 6	SLK 7	SLK 8	SLK 9	SLK 10	SLK 11	SLK 12
1	35.000	35.000	0.000	0.000	0.000	0.000	0.000
2	0.000	0.000	0.000	0.000	0.000	0.000	0.000
3	1.000	1.000	0.000	0.000	0.000	0.000	0.000
4	1.000	1.000	0.000	0.000	0.000	0.000	0.000
5	-1.000	-1.000	0.000	0.000	0.000	0.000	0.000
6	-1.000	0.000	0.000	0.000	0.000	0.000	0.000
7	0.000	-1.000	0.000	0.000	0.000	0.000	0.000
8	-0.100	-0.100	1.000	0.000	0.000	0.000	0.000
9	-0.500	0.000	0.000	1.000	0.000	0.000	0.000
10	-0.200	0.000	0.000	0.000	1.000	0.000	0.000
11	0.000	-0.100	0.000	0.000	0.000	1.000	0.000
12	0.000	-0.100	0.000	0.000	0.000	0.000	1.000
13	-0.100	-0.100	0.000	0.000	0.000	0.000	0.000

ROW		SLK 13
1	0.00E+00	-0.19E+07
2	0.000	15000.000
3	0.000	0.000
4	0.000	10000.000
5	0.000	14000.000
6	0.000	7000.000
7	0.000	25000.000
8	0.000	700.000
9	0.000	3500.000
10	0.000	1400.000
11	0.000	2500.000
12	0.000	2500.000
13	1.000	420.000

LP OPTIMUM FOUND AT STEP 7

OBJECTIVE FUNCTION VALUE

1)	1904000.	
VARIABLE	VALUE	REDUCED COST
P1	0.000000	0.000000
R1	15000.000000	0.000000
E1	0.000000	0.000000
P2	0.000000	0.000000
R2	7000.000000	0.000000
E2	25000.000000	0.000000
P3	14000.000000	0.000000
R3	0.000000	0.000000
E3	0.000000	0.000000

ROW	SLACK OR SURPLUS	DUAL PRICES
2)	0.000000	9.000000
3)	0.000000	3.000000
4)	10000.000000	0.000000
5)	0.000000	-35.000000
6)	0.000000	-35.000000
7)	0.000000	-35.000000
8)	700.000000	0.000000
9)	3500.000000	0.000000
10)	1400.000000	0.000000
11)	2500.000000	0.000000
12)	2500.000000	0.000000
13)	420.000000	0.000000

NO. ITERATIONS= 7

RANGES IN WHICH THE BASIS IS UNCHANGED:

OBJ COEFFICIENT RANGES

VARIABLE	CURRENT COEF	ALLOWABLE INCREASE	ALLOWABLE DECREASE
P1	26.000000	INFINITY	0.000000
R1	26.000000	0.000000	INFINITY
E1	26.000000	INFINITY	0.000000
P2	32.000000	0.000000	0.000000
R2	32.000000	0.000000	0.000000
E2	32.000000	0.000000	35.000000
P3	35.000000	0.000000	3.000000
R3	35.000000	INFINITY	0.000000
E3	35.000000	INFINITY	0.000000

RIGHTHAND SIDE RANGES

ROW	CURRENT RHS	ALLOWABLE INCREASE	ALLOWABLE DECREASE
2	15000.000000	4200.000000	0.000000
3	32000.000000	4200.000000	0.000000
4	24000.000000	INFINITY	10000.000000
5	14000.000000	10000.000000	14000.000000
6	22000.000000	0.000000	4200.000000
7	25000.000000	0.000000	4200.000000
8	0.000000	700.000000	INFINITY
9	0.000000	3500.000000	INFINITY
10	0.000000	INFINITY	1400.000000
11	0.000000	2500.000000	INFINITY
12	0.000000	INFINITY	2500.000000
13	0.000000	INFINITY	420.000000

THE TABLEAU

ROW	(BASIS)		P1	R1	E1	P2	R2	E2	
1		ART	0.000	0.000	0.000	0.000	0.000	0.000	
2	R1		1.000	1.000	1.000	0.000	0.000	0.000	
3	P2		1.000	0.000	0.000	1.000	0.000	0.000	
4		SLK		4	0.000	0.000	0.000	0.000	0.000
5	P3		0.000	0.000	0.000	0.000	0.000	0.000	
6	R2		-1.000	0.000	-1.000	0.000	1.000	0.000	
7	E2		0.000	0.000	1.000	0.000	0.000	1.000	
8		SLK		8	0.150	0.000 0.000	0.000	0.000	
9		SLK		9	-0.500	0.000	-0.500	0.000	0.000
10		SLK		10	-0.200	0.000	-0.200	0.000	0.000
11		SLK		11	0.000	0.000	0.150	0.000	0.000
12		SLK		12	0.000	0.000	0.200	0.000	0.000
13		SLK		13	-0.050	0.000	0.000	0.000	0.000

ROW	P3	R3	E3	SLK 2	SLK 3	SLK 4	SLK 5
1	0.000	0.000	0.000	9.000	3.000	0.000	35.000
2	0.000	0.000	0.000	1.000	0.000	0.000	0.000
3	0.000	-1.000	-1.000	1.000	1.000	0.000	0.000
4	0.000	0.000	0.000	1.000	1.000	1.000	1.000
5	1.000	1.000	1.000	-1.000	-1.000	0.000	-1.000
6	0.000	1.000	0.000	-1.000	0.000	0.000	0.000
7	0.000	0.000	1.000	0.000	0.000	0.000	0.000
8	0.000	0.100	0.100	-0.100	-0.100	0.000	-0.050
9	0.000	1.000	0.000	-0.500	0.000	0.000	0.000
10	0.000	0.000	0.000	-0.200	0.000	0.000	0.000
11	0.000	0.000	-0.100	0.000	0.000	0.000	0.000
12	0.000	0.000	0.000	0.000	0.000	0.000	0.000
13	0.000	0.100	0.100	-0.100	-0.100	0.000	-0.030

ROW	SLK 6	SLK 7	SLK 8	SLK 9	SLK 10	SLK 11	SLK 12
1	35.000	35.000	0.000	0.000	0.000	0.000	0.000
2	0.000	0.000	0.000	0.000	0.000	0.000	0.000
3	1.000	1.000	0.000	0.000	0.000	0.000	0.000
4	1.000	1.000	0.000	0.000	0.000	0.000	0.000
5	-1.000	-1.000	0.000	0.000	0.000	0.000	0.000
6	-1.000	0.000	0.000	0.000	0.000	0.000	0.000
7	0.000	-1.000	0.000	0.000	0.000	0.000	0.000
8	-0.100	-0.100	1.000	0.000	0.000	0.000	0.000
9	-0.500	0.000	0.000	1.000	0.000	0.000	0.000
10	-0.200	0.000	0.000	0.000	1.000	0.000	0.000
11	0.000	-0.100	0.000	0.000	0.000	1.000	0.000
12	0.000	-0.100	0.000	0.000	0.000	0.000	1.000
13	-0.100	-0.100	0.000	0.000	0.000	0.000	0.000

ROW	SLK 13	
1	0.00E+00	-0.19E+07
2	0.000	15000.000
3	0.000	0.000
4	0.000	10000.000
5	0.000	14000.000
6	0.000	7000.000
7	0.000	25000.000
8	0.000	700.000
9	0.000	3500.000
10	0.000	1400.000
11	0.000	2500.000
12	0.000	2500.000
13	1.000	420.000

8. Data envelopment analysis and linear programming

8.1. Modeling of ranking units using Data Envelopment Analysis (DEA)

Data envelopment analysis (DEA), occasionally called frontier analysis, was first put forward by Charnes, Cooper and Rhodes in 1978. It is a performance measurement technique which, as we shall see, can be used for evaluating the *relative efficiency* of *decision-making units (DMU's)* in organizations. Here a DMU is a distinct unit within an organization that has flexibility with respect to some of the decisions it makes, but not necessarily complete freedom with respect to these decisions.

Examples of such units to which DEA has been applied are: banks, police stations, hospitals, tax offices, prisons, defense bases (army, navy, air force), schools and university departments. Note here that one advantage of DEA is that it can be applied to non-profit making organizations.

Since the technique was first proposed much theoretical and empirical work has been done. Many studies have been published dealing with applying DEA in real-world situations. Obviously there are many more unpublished studies, e.g. done internally by companies or by external consultants.

We will initially illustrate DEA by means of a small example. More about DEA can be found on line using Google: "Data Envelopment Analysis". Note here that much of what you will see below is a graphical (pictorial) approach to DEA. This is very useful if you are attempting to explain DEA to those less technically qualified (such as many you might meet in the military or management world). There is a mathematical approach to DEA that can be adopted however. We will present the single measure first to demonstrate the idea and then move to multiple measures and use linear programming methodology from our course.

Example 1. Ranking Banks

Consider a number of bank branches. For each branch we have a single output measure (number of personal transactions completed) and a single input measure (number of staff).

The data we have is as follows:

Branch	Personal transactions ('000s)	Number of staff
Branch 1	125	18
Branch 2	44	16
Branch 3	80	17
Branch 4	23	11

For example, for the Branch 2 in one year, there were 44,000 transactions relating to personal accounts and 16 staff members were employed.

How then can we compare these branches and measure their performance using this data?

Ratios

A commonly used method is *ratios*. Typically we take some output measure and divide it by some input measure. Note the terminology here, we view branches as taking *inputs* and converting them (with varying degrees of efficiency, as we shall see below) into *outputs*.

For our bank branch example we have a single input measure, the number of staff, and a single output measure, the number of personal transactions. Hence we have:

Branch	Personal transactions per staff member ('000s)
Branch 1	6.94
Branch 2	2.75
Branch 3	4.71
Branch 4	2.09

Here we can see that Branch1 has the highest ratio of personal transactions per staff member, whereas Branch 4 has the lowest ratio of personal transactions per staff member.

As Branch 1 has the highest ratio of 6.94 we can compare all other branches to it and calculate their *relative efficiency* with respect to Branch 1. To do this we divide the ratio for any branch by 6.94 (the value for Croydon) and multiply by 100 to convert to a percentage. This gives:

Branch	Relative efficiency
Branch 1	100(6.94/6.94) = 100%
Branch 2	100(2.75/6.94) = 40%
Branch 3	100(4.71/6.94) = 68%
Branch 4	100(2.09/6.94) = 30%

The other branches do not compare well with Branch 1, so are presumably performing less well. That is, they are relatively less efficient at using their given input resource (staff members) to produce output (number of personal transactions).

We could, if we wish, use this comparison with Branch 1 to set *targets* for the other branches.

For example we could set a target for Branch 4 of continuing to process the same level of output but with one less member of staff. This is an example of an *input target* as it deals with an input measure.

An example of an *output target* would be for Branch 4 to increase the number of personal transactions by 10% (e.g. by obtaining new accounts).

Plainly, in practice, we might well set a branch a mix of input and output targets which we want it to achieve. We can use linear programming.

8.2. Linear Programming example of DEA

Example 2. Ranking banks with linear programming

Typically we have more than one input and one output. For the bank branch example suppose now that we have two output measures (number of personal transactions completed and number of business transactions completed) and the same single input measure (number of staff) as before.

The data we have is as follows:

Branch	Personal transactions ('000s)	Business transactions ('000s)	Number of staff
Branch 1	125	50	18
Branch 2	44	20	16
Branch 3	80	55	17
Branch 4	23	12	11

We start be scaling (via ratios) the inputs and outputs to reflect the ratio of 1 unit.

Branch	Personal transactions ('000s)	Business transactions ('000s)	Per employee or staff
Branch 1	125/18=6.94	50/18=2.78	18/18=1
Branch 2	44/16=2.75	20/16=1.25	16/16=1
Branch 3	80/17=4.71	55/17=3.24	17/17=1
Branch 4	23/11=2.09	12/11=1.09	11/11=1

Pick a DMU to maximize: $E1$, $E2$, $E3$, or $E4$

Let $W1$ and $W2$ be the personal and business transactions at branch

In this example we choose to maximize branch two, $E2$.

Here is the LP formulation for this DEA problem:

Maximize $E2$

Subject to

$E1 = 6.94\,W1 + 2.78\,W2$
$E2 = 2.75\,W1 + 1.25\,W2$
$E3 = 4.71\,W1 + 3.24\,W2$
$E4 = 2.09\,W1 + 1.09\,W2$
$E1 \leq 1$
$E2 \leq 1$
$E3 \leq 1$
$E4 \leq 1$

Now, what did we learn from this. If we ranked ordered the branches on efficiency performance of our inputs and outputs, we find

Branch 1	100%
Branch 3	100%
Branch 2	43.2%
Branch 4	36.2%

We know we need to improve on branch 2 and branch 4 performances while not losing our efficiency in branches 1 and 3. A better interpretation could be that the practices and procedures used by the other branches were to be adopted by Branch 4, they could improve their performance.

This invokes issues of highlighting and disseminating examples of best practices. Equally there are issues relating to identification of poor practices.

In DEA the concept of the reference set can be used to identify best performing branches with which to compare poorly performing branches. If you use this procedure, use it wisely.

Author details

William P. Fox[1] and Fausto P. Garcia[2]

1 Naval Postgraduate School, USA

2 Universidad Castilla-La Mancha, Spain

References

[1] Apaiah, R. & E. Hendrix ((2006). Linear programming for supply chain design: A case on Novel protein foods. Ph.D. Thesis, Wageningen University, Netherlands.

[2] Balakrishnan, N. B, & Render, R. Stair. ((2007). Managerial Decision Making, 2nd Ed. Saddle River, NJ: Prentice Hall.

[3] Bazarra Mokhtar S. J.J. Jarvis, and H. D. SheralliLinear Programming and Network Flows, New York. John Wiley & Sons, (1990).

[4] EckerJoseph and M. Kupperschmid, Introduction to Operations Research, John Wiley and Sons, (1988).

[5] Fox, W. P. (2012). Mathematical Modeling with Maple. Boston, MA: Cengage Publishers.

[6] Giordano, F, Fox, W, Horton, S, & Weir, M. (2009). A First Course in Mathematical Modeling, 4th Ed. Belmont, CA: Brooks-Cole.

[7] Hiller Fredrick S and Gerald J. LibermanIntroduction to Mathematical Programming, McGraw Hill Publishing Company, (1990).

[8] Winston Wayne L. Introduction to Mathematical Programming Applications and Algorithms, 4th Edition. Belmont, CA. Duxbury Press. (2002).

Chapter 9

The Investment in Hedge Funds as an Alternative Investment

Joaquín López Pascual

Additional information is available at the end of the chapter

1. Introduction

One of the most relevant industries of the modern financial management is the investment in hedge funds. The hedge fund industry is a heterogeneous group. One way to classify hedge funds is according to the investment strategy used, each offering a different degree of return and risk. Their historical return distributions provide with key information in order to understand the strategies behaviour.

In this Chapter, we briefly introduce the basic concepts about investment in hedge funds. We will provide some general definitions and introduction to hedge funds, their analysis and strategies. For a more extensive guide over hedge funds analysis please refer to López Pascual, J and Cuellar, R.D. (2010):"Assessment and Selection of Hedge Funds and Funds of Hedge Funds" Working Paper Nº5 2010, Cunef.

We present here the problems related to risk analysis in hedge funds and problems related to accounting or valuation of illiquid assets. Moreover, we introduce the various most common hedge fund strategies and their expected risks. Can we reach a market neutral investment strategy by investing in hedge funds? How can we build a diversified portfolio of hedge funds? How can we measure risk adjusted returns and total risk in hedge funds? These are some of the questions to which we aim to respond in this Chapter.

Finally, we will show inherent characteristics of different hedge fund strategies and illustrate how popular analysis tools such as; Sharpe ratio, Sortino ratio or other techniques, that take high moments as inputs, systematically overestimate or underestimate the risks of certain strategies. This corroborates our point that manager selection has to be contextualized according to the strategy employed

One of the most important attributes of a hedge fund is the ability to perform above a certain hurdle rate at all times no matter what market conditions prevail. This attribute has been called market neutral, which under no circumstances should be considered as neutral to the markets. As the LTCM (Long Term Capital Management) experience has demonstrated, there is no hedge fund that can be completely unaffected by a general adverse prevailing market condition. However, some managers are able to turn an adverse market condition into an opportunity, delivering extraordinary returns during market turmoil.

Understanding hedge funds is not a very easy task. There are a number of complexities involved in investments related to hedge funds. Legal and compliance, operations, qualitative analysis and quantitative analysis, and technology related questions means that operational due diligence is a very important concept in the allocation to hedge funds.

In general, it is considered that hedge funds have to be beta neutral or that the level of correlation with the performance benchmark of the market where the fund is involved should be as close to zero as possible. The principal function of the hedge fund in this conceptual frame would be of at least capital preservation in bear markets and capital appreciation in bullish markets.

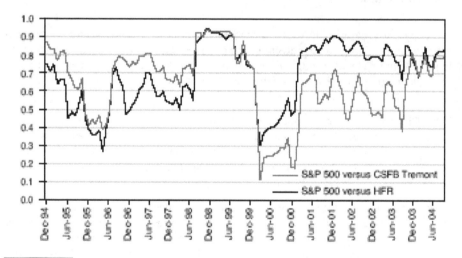

Source: Bloomberg

Figure 1. Rolling 12 month correlation between the S&P 500 and the HFR Equity Hedge Index and the CSFB Tremont Long/Short Equity Index from the period 1993-2004

This definition calls for reviewing the concept of absolute returns, which have been in the area of investment since the inception of hedge funds into the arena of investment vehicles. Recent research (Waring and Siegel, 2006) explores the frontiers of alpha generation. It is considered that a portfolio manager is exposed to beta, but returns exceeding beta exposure can be attributed to manager's skills as measured by the alpha. However, as already mentioned, hedge fund managers are not always able to generate alpha and they are even sometimes not able to beat passively managed investment portfolios, such as index funds, which does not necessarily mean that they are not alpha generators given their non-directional investment style. As we know, beta can be obtained in the market to significantly cheaper prices than hedge fund fees, just by investing in an index replicating an investment portfolio or by using derivatives or, most recently ETFs, which are very liquid actively managed instruments and are able to provide a number of products for beta generation. Beyond the beta, the most important aspect in hedge fund selection is the manager's abilities to generate returns by his skills because, as demonstrated by the research mentioned above, there is no such thing as an absolute-return investor, but a relative return investor. It is correct to assess that a well managed hedge fund is one that has a zero or nearly close to zero beta coefficient, as we can observe in the Figure 1., while enjoying a high degree of alpha in its portfolio returns.

The question is how an investor can be able to assess the level of alpha generation by a hedge fund manager. Analysing the track record of the fund is a possible answer. However, in doing so, investors should be aware that historical performance is not a guarantee of future returns. The consistencies between historical and future returns have to be carefully assessed considering a number of parameters that result in higher and consistent alpha creation. However, one should consider that linear factor models such as the ones developed by Markowitz or Sharpe are unable to capture hedge fund's nonlinear return features.

In line with this assessment, Fung and Hsie (2001) have developed a model based on asset-based style factors. These factors with statistical significance may not necessarily be associated to any strategy or specific investment style. The statistical clustering created by using principal component analysis (PCA) is able to group common risk and returns characteristics of the sample. This is very important because hedge funds are actively managed investment organizations, so timing and leverage are relevant influential factors of the investment style and strategy. The attractivity of the non-correlated returns generated by hedge funds bearing low or neutral beta and a high alpha should be assessed in the context of portfolio diversification. Kat (2005) established that the undesired effects of hedge funds that are attributable to negative skewness and high kurtosis can be eliminated through the use of out-of-the-money put options or by investing in other hedging strategies. In this context, it is clear that hedge fund returns are not "superior", but different than returns generated by other asset classes. Needless to mention, a diversified hedge fund portfolio has for a retail investor a prohibitive cost, given the fact that the minimal investment in an average hedge fund is in the order of USD 1 million and a diversified portfolio should have about 10 to 15 underlying vehicles.

The search for superior, and uncorrelated, returns leads investors to seek "alternative" investments. This term is certainly not precise but a high level of interest has been concentrated on the hedge fund industry as a paradigm of alternative investments; however this asset class is a heterogeneous group.

One way to classify hedge funds is according to the investment strategies employed. The strategies perform differently according to the economic cycle, each offering a different degree of return and risk. Therefore performance, generated in a specific part of an economic cycle, that seem to have achieved consistent high excess returns could underperform systematically once that the economic cycle changes.

The returns generated by a hedge fund have to be understood in the context of the strategy used and the economic cycle. This implies a double problem for investors:

- The allocation strategy will depend on what investors are looking for. For instance, are they looking for a dynamic hedge to the equity market? In this case a negative beta strategy as "short bias" could be the right decision. It will be more complicated if the investors looks for absolute returns. In this case, the first decision is to decide between a passive or a dynamic approach. Another complicated choice would be a hedge fund that enhances portfolio efficiency.

- In the quest for the right hedge fund, a key factor to understand is the intrinsic behaviour of the strategy followed by the manager. Then, the investors have to look for a manager with an "edge" in the specific strategy.

2. Definitions

2.1. What is a hedge fund?

There is no an universally accepted definition of *hedge fund*. However, the common characteristics of the term *hedge fund* are; private investment fund that invest in a wide range of assets and employs a great variety of investment strategies. Due to their nature hedge funds have almost no restrictions in the use of derivatives, leverage or short-selling. This combination of capacity, instruments and flexibility in their investment decisions makes a significant difference relative to the more regulated, mutual funds.

Also, the combination of these resources has allowed hedge funds to exploit new market opportunities through investment strategies.

2.2. Investment strategies

There is also no consensus regarding the number of investment strategies used by hedge funds. Financial technology evolves and the universe of investment assets is constantly growing. Therefore new investment strategies are continually developing to exploit market opportunities. Even hedge funds that invest in the same type of assets can try to make money taking exposure to different risk factors. For example a hedge fund investing in convertible bonds could be aiming to get equity, credit, volatility, liquidity, interest rate exposure, or a combination of several of them. The exposure to each of these factors could be exploited through different strategies. Therefore, it is important to note that different strategies provide a different degree of return and risk.

2.3. Hedge fund indexes

Hedge funds have no formal obligation to disclose their results, however most of the funds release, at least monthly, their returns to attract new investors. With this information some data vendors have built performance hedge fund indexes, as well as sub indexes according to the fund strategy.

Some of these data providers are;

• Hedge Fund Research (HFR)

• Zurich Capital Markets

• CSFB Tremont

• Hennesse

• Tuna

• Barclays

2.4. Historical return analysis

The historical return analysis provides an important source of information for evaluating and understanding hedge funds investment styles.

Through explicit or implicit analysis we can try to explain the funds performances and to classify investment styles.

• *Explicit analysis.* The aim is to identify and measure the sensitivity of real factors that explain the historical returns. An example could be to model the returns as a linear function of various macro economic factors or indexes.

• *Implicit analysis.* The idea is to identify certain statistical factors that explain the historical returns. One the most used methods is the principal component analysis (PCA). The PCA ranks explanatory factors with the highest possible variance with the constraint that each one has to be orthogonal to the previous components.

In addition, comparing the time series returns of a hedge fund against the returns of its peer group will allow us to assess the investment skills of the manager.

From an investor's perspective, it is important to maintain a clear view of the risk exposure gained by a hedge fund investment in relation not only to the returns but also with the investment vehicle strategy. Different strategies yield not only different risk exposures but expose the investment to different risk classes. In this respect, it is important to conceptualize the risk. Some investors wrongly believe that by investing in bonds or in an investment fund, which invests in fixed income securities, they are only exposed to interest rate risk or credit risk. A brief list of possible risks that investors face in financial markets can be summarised as follows:

Accounting risk	Fiduciary risk	Political risk
Bankruptcy risk	Hedging risk	Prepayment risk
Basis risk	Horizon risk	Publicity risk
Call risk	Iceberg risk	Regulatory risk
Capital risk	Interest rate risk	Reinvestment risk
Collateral risk	Knowledge risk	Rollover risk
Commodity risk	Legal risk	Spread risk
Concentration risk	Limit risk	Systemic risk
Contract risk	Liquidity risk	Taxation risk
Currency risk	Market risk	Technology risk
Curve construction risk	Maverick risk	Time lag risk
Daylight risk	Modeling risk	Volatility risk
Equity risk	Netting risk	Yield curve risk
Extrapolation risk	Optional risk	

Source: Author based on CMRA

Figure 2. Partial listing of risk universe in relation to hedge funds

Institutional investors have traditionally used asset allocation as the core process in order to determine their investment strategy. The process of asset allocation is important; however, it does not take into account the dynamic changes in risk appetite and the changing dynamics of risk in the investment portfolio. Risk budget monitoring introduces a different dimension in the investment process as a function of volatility, correlation, and investment volume itself.

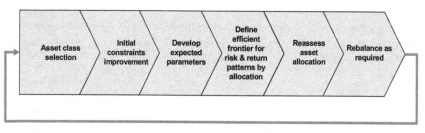

Source: Author

Figure 3. Asset allocation process

Risk budgeting is a tool that should not be seen as an optimization process, because the optimization process in asset allocation uses a traditional mean-variance approach to efficiently allocate assets in a trade-off process of risk and returns. The objective of optimal investment risk management has to be such that it allows the investor to acquire less risk for a larger return or more return in exchange for the current risk exposure.

Other than the universe of possible risks mentioned in Figure 2, hedge funds gain exposure through poor market liquidity, use of leverage, high turnover, heavy use of derivatives instruments, correlation to unrelated assets and transparency risk, to mention just a few. Risk measurement in traditional investment vehicles or asset classes seems to be a very straight-forward exercise when compared with hedge funds.

Asset allocation is concerned with optimal asset combination, thus mathematically it is equivalent to a constrained optimization process. The process of asset allocation is much simpler than portfolio construction. Brinson et al. (1986, 1991) established that more than 90 percent of the variability of an investment portfolio is due to asset allocation. The advantage of the asset allocation process is that we resolve the optimization process at the asset class level instead of at the single security level. This is simpler because it is easier to estimate expected future returns at the asset class level than at the single security level and because the correlations are clearly established in order to build a diversified allocation. In this frame, we should consider investment in alternative funds as an asset class problem within the optimization process of asset allocation. Empirical research (Lintner, 1983) has robustly established the virtues of including alternative assets in the allocation process given the low and even negative correlations with traditional asset classes.

One of the main challenges for investors is the poor transparency of hedge funds, which allows for very important risk misspecifications. The non-stationarity of risk due to the dynamic asset allocation of hedge fund managers is another challenge in risk measurement. Under these circumstances, it is very difficult to reduce measurement error to near zero. Identifying risk in a dynamic investment environment requires high frequency assessment and great accuracy. Factor analysis can not only assist in identifying risk factors but also the rate of change of those factors. Factor analysis can determine the aggregate factors explaining investment returns. This analysis can be used either as forward risk modelling or as inverse modelling.

Forward risk modelling uses assumed pre-existing risk factors to assess the risk universe of the investment portfolio. If the investor has allocated investments to hedge funds using a convertible arbitrage strategy, we can assume risk factors correlated to fixed income securities as well as stocks, because such an investment strategy is exposed not only to risk factors related to the yield curve but also because when the hedge fund manager exercises his option in a convertible bond, he is automatically gaining exposure to stock market risks.

Static forward modelling (SFM) analyses the returns and finds the factors that can fit in the return's model. By definition, SFM is a replication strategy using future contracts or other trading assets. The modelling eliminates sequentially uncorrelated factors that assist in explaining the stream of returns. In practice, SFM is used as an early warning system for the fund of funds manager, because when the manager sees a new factor emerging which can affect the returns directly or indirectly, the manager should try to rebalance the portfolio eliminating the style drifting underlying position.

Inverse risk modelling uses principal component analysis (PCA) in order to analyse time series of returns and establishes all possible patterns with exposure to risk factors explaining the returns. Using the covariance matrix, the manager extracts the eigenvectors with maximum explanatory power in statistical terms, but because these eigenvectors are not the real economic variables such as actual gold price or the exchange parity of currencies, the manager must correlate the characteristics of those statistical factors to real factors. Interpretation is in this case absolutely critical but many times is not even possible.

Non-stationary or dynamic factor analysis takes into consideration relative changes of exposure along a time series of factors or combination of factors and their weights in explaining the returns of a portfolio. Managers have to take into account a sufficiently long horizon that explains the trade-off between risk and returns. When the factors and the returns converge in a time series, there is an alignment in the risk factors and the established strategy. Observation has to be maintained for a certain period of time because at a certain point the exposures could be subject to variations and diversions, letting the manager without knowledge of the new risk factors. The use of multi-scale correlation methods can assist portfolio managers in establishing the right time horizon for the analysis. Two significant risks in the analysis can be found. The first is that the time horizon of the assessment is too short and the point of divergence between the explaining factors and the portfolio return streams cannot be evaluated with a certain degree of accuracy, and the second is that the established time horizon is too long diluting the effects so much that the factors combination and the moment relation can hardly be visualized.

Detecting changes in correlations or non-stationarities across time is very useful for the investor because with assistance of this multi-scale correlation method, we can build an error map. If the error map becomes non-zero, it is because the correlation between the explaining factors and the returns has collapsed. Collapse in return attributions are a warning indicator that the fund manager has changed the strategy or is entering into a strategy shifting process that should trigger an immediate explanation by the fund manager to the investor about this change and the new risk factors implicated in such a strategy move. Another indicator of strategy shifting is sudden factor dispersion, which is given through introduction of new explanatory factors or alterations in the eigenvectors of the covariance matrix, which again we insist are statistical factors that have to be correlated to real economically relevant factors such as interest rate risk, volatility index (VIX), or gold price, to mention a few examples.

The practice of investment portfolio risk budgeting in the context of hedge fund management is to align risk budgeting with a coherent risk measurement methodology in order to obtain an appropriate risk amount. There are a number of variations of VaR methodologies of which the most utilized in the hedge fund industry is certainly CVaR (Rockefellar and Uryasev, 2000 and 2001).

A key factor to successful risk management in the context of hedge fund investment management and monitoring is to include a stress test in the risk to be budgeted and allocated. In this context, it is important to remember that VaR does not capture all the essence of risk in hedge funds. An example is that VaR has failed to capture the risks of instability related to the euro-convergence during the 1990s. Stress analysis instead tries to resolve questions such as:

1. Which variables, given a certain variation, affect and to what degree the price of an asset

2. Which are the variables, given a rate of change, can affect the valuation of the portfolio and to what extend and for how long

3. How wide is the variance established by the fund manager for the relevant variables affecting the portfolio and how these divert from other portfolio managers

4. How accepted and valid is the approach used by the portfolio manager compared to other peers

Stress test results need to be integrated into the denominator of the risk adjusted reward equation. In stress testing results should be included not only variations in market moves but also assumptions underlying strategies, as well as the possible adverse effects on the portfolio of liquidity premium, on-the-run and off-the run differential credit spread sensitivities, haircut sensitivity, and sensitivity to correlations. Investment managers control risks by closely monitoring the variety and level of exposures to different risk categories. In hedge fund management, one of the most critical risks is liquidity risk. Managers and investors need to understand that valuing positions at mid-market when positions are large and market liquidity is poor can be very misleading. A natural reflex in market turmoil scenarios is always to liquidate the most liquid instruments in the portfolio to meet margin demands by prime brokerages. As we have seen in the case of LTCM, this is normally the equivalent of a death sentence because it constrains the portfolio to the most illiquid instruments leaving the managers in a very vulnerable position in a distressed market.

Different hedge fund strategies deliver not only different returns but also different risk exposures. Investors investing in a portfolio of hedge funds should visualize clearly their exposures and the level of concentration to those at any given time. As mentioned, there are a variety of hedge fund strategies that give investors different exposures to different risks. Lhabitant (2004) uses an adapted version of the Herfindahl-Hirschman index to assess the level of concentration to certain strategies by an investor:

In this case, the investor clearly understands the normalized sum of squared styles concentration. Moreover, as we know, investors have to visualize their exposure to different sets of risks, which are implied in each hedge fund strategy. Different data vendors providing style benchmarks have classified different hedge fund strategies, as described in Figure 4.

In this Chapter, we consider that the one of the most reliable data sources is the one provided by Credit Suisse Tremont, Greenwich Alternative Investment (former Van Hedge), Hedge Fund Research, and Barclay Hedge Fund Index.

Moreover, EDHEC Business School has made an index of indexes using PCA in order to homogenize the strategy universe of hedge funds. Based on these strategies and considering the particularities from each hedge fund, investors can use the strategy definitions by the data vendors and correlate every strategy with the typical or specific risks factors of each strategy and according to the operational due diligence performed on the fund strategy.

The fundamental aspects to consider are the visualization of risk, qualification, and quantification of risk exposure. As described before, with the assistance of PCA investors can evaluate the relevant risk factors related to the fund's strategy and then correlate them with real economic risk factors.

Credit Suisse Tremont	Greenwich Alternative Investments	Hedge Fund Research	Barclay Hedge Fund Index
Convertible Arbitrage	Equity Market Neutral	Convertible Arbitrage	Convertible Arbitrage
Dedicated Short-Bias	Event Driven	Distressed Securities	Distressed Securities
Emerging Markets	Distressed Securities	Equity Hedge	Emerging Markets
Market neutral	Merger Arbitrage	Equity Market Neutral	Equity Long Bias
Event Driven:	Special Situations	Event Driven	Equity Long/Short
Distressed	Market Neutral Arbitrage	Macro	Equity Market Neutral
Multi-Strategy	Convertible Arbitrage	Merger Arbitrage	Equity Short Bias
Risk Arbitrage	Fixed Income Arbitrage	Relative Value Arbitrage	European Equities
Fixed Income Arbitrage	Other Arbitrage		Event Driven
Global Macro	Statistical Arbitrage		Fixed Income Arbitrage
Long/Short Equity	Aggresive Growth		Fund of Funds
Managed Futures	Opportunistic		Global Macro
Multi-Strategy	Short Selling		Health Care & Biotechnology
	Value		Merger Arbitrage
	Futures		Multi-Strategy
	Macro		Pacific Rim Equities
	Market Timing		Technology
	Emerging Markets		
	Income		
	Multi-Strategy		

Source: Author

Figure 4. Different strategies according to different hedge fund index providers

Investors should consider that short positions are always at risk of liquidity squeeze. This kind of risk is entered when short positions are negatively affected by market prices development, generating potential or real losses in the portfolio and forcing the prime broker to place margin calls to increase collateral in form of cash or securities to cover for possible or effective losses or when the prime broker calls the loaned securities, forcing the fund manager to generate losses from the positions. This is the case when the prices of the short positions are rising above the collateral held by the prime broker in form of cash or cash equivalents. Short positions normally act as a hedging in a long portfolio segment instead or as a complement to derivative instruments, reducing significantly the cost of hedging but exposing the portfolio to its own set of risks.

Since is very difficult to generate ideas in bull markets about possible losers, some hedge funds either outsource to other funds the shorts or they hedge their position with the use of option derivatives of all sorts, which by all means is a more sophisticated hedging but sometimes a very expensive one. In general, it has been established in hedge funds to hedge long positions by shortening others.

3. Investment strategies and indexes

We can use some of the HFR sub indexes as proxies for hedge fund strategies performance.

HFR have developed a series of benchmark indexes designed to reflect hedge fund industry performance by constructing equally weighted composites of constituent funds. The funds selected in each index are filtered through manager's due diligence and other qualitative requirements. The classification is done through statistical analysis, cluster analysis, correlation analysis, optimization and Monte Carlo simulations. This information is available on the HFR web site.

One of the main problems with the hedge fund indexes is the survivorship bias. Many hedge funds that were included at some point in the indexes might now not comply with the index requirements or might be defunct. HFR minimizes this problem by trying to receive a fund's performance until the point of the final liquidation of the fund.

HFR has created the following index classification.

Hedge Fund Strategy Classifications			
Equity Hedge	Event Driven	Macro	Relative Value
Equity Market Neutral	Activist	Active Trading	Fixed Income - Asset Backed
Fundamental Growth	Credit Arbitrage	Commodity: Agriculture	Fixed Income - Convertible Arbitrage
Fundamental Value	Distressed / Restructuring	Commodity: Energy	Fixed Income - Corporate
Quantitative Directional	Merger Arbitrage	Commodity: Metals	Fixed Income - Sovereign
Sector: Energy/Basic Materials	Private Issue/ Regulation D	Commodity: Multi	Volatility
Sector: Technology/Healthcare	Special Situations	Currency: Discretionary	Yield Alternatives: Energy Infrastructure
Short Bias	Multi-Strategy	Currency: Systematic	Yield Alternatives: Real Estate
Multi-Strategy		Discretionary Thematic	Multi-Strategy
		Systematic Diversified Multi-Strategy	

Source: www.hedgefundresearch.com

Table 1. Add caption

Based on these indexes we could show inherent characteristics of some of the most relevant hedge fund strategies and we would assimilate the historical returns of each investment strategy taking a long or short positions in plain vanilla options.

4. Return distributions and ratios

Most of the strategies, except *Short Bias*, show common characteristics as negative skewness, positive excess kurtosis and serial correlation.

The main consequence of these characteristics is that left tail of the return distribution is longer than the right side; therefore large losses are bigger than the suggested by the standard deviation. Furthermore, the serial correlation of the returns hide that the model underestimates the true variance and reduce the effective number of degrees of freedom in a time series. In the case of hedge funds analysis, it means that we will be, underestimating the true risk of our investment and, over allocating to hedge funds when we undertake a mean variance portfolio analysis.

4.1. Effect of the serial correlation in a distribution

Brooks and Kat (2002) argued that the serial correlation of the hedge funds returns seems inconsistent with the notion of efficient markets. According with them, one possible explanation could be the fact that many hedge funds invest in illiquid or complex assets. To find update valuations of these assets is not always an easy task; therefore sometimes they use the last reported transaction price or model valuations. López and Cuellar (2007), explained the hedge fund returns serial correlation with similar arguments, appointing that real state valuations show the same problem due to the illiquid securities to appraise. These explanations are also consistent with Agarwal, V., Daniel, N.D and Naik, N.Y findings. They found that hedge funds, up to a certain extent, manage the reported returns in order to "smooth" their return distributions.

4.2. Ratios

The analysis of hedge funds performances through ratios is an easy and intuitive way to measure the efficiency of an investment. López de Prado (2008) appoints that the Sharpe ratio has become the 'gold standard' of performance evaluation. Although many researchers, Sharpe himself, study the deficiencies and limitations of the ratio, rating agencies and institutional investors include this ratio in their performance and risk measurements as appointed López and Cuellar (2007).

The two most used ratios are; Sharpe and Sortino, both measure the excess returns of an investment per unit of risk. In the case of the Sharpe ratio, the unit of risk is calculated as the standard deviation of the investment returns. For the Sortino ratio, the unit of risk is measure as the standard deviation of the negative returns. In other words, a measure of excess return against downward price volatility.

The statistical characteristics of the hedge funds returns result in overestimated Sharpe or Sortino ratios, therefore these ratios tend to overvalue the efficiency of hedge funds and drive to over allocation in this asset class. This technology has limitations therefore the results have to be understood in the context of the selected strategy and the inherent risks. In this direction, López and Cuellar (2010) propose a complementary system for evaluating the inherent risks of each hedge fund through a radar visualization of strategy exposure.

5. Conclusions

Understanding the statistical behaviour of hedge fund strategies is a key factor in order to select hedge fund investments. Study of their historical returns will provide us with a lot of information; however it is important to understand the limitations of the technology used. Performances generated in a specific part of an economic cycle, that seem to have achieved consistent high excess returns could underperforms systematically once that the economic cycle changes, therefore the returns generated by a hedge fund has to be understood in the context of the strategy used and the economic cycle.

Our conclusion is reached for the hedge fund industry as a whole; this conclusion is not in conflict with the fact that many hedge fund managers consistently attain returns for their investors that amply justify the fees charged. However the fact that certain hedge fund managers are worth the current structure of fees is not a justification for the fees charged across the whole industry.

Author details

Joaquín López Pascual

Address all correspondence to: joaquinlopez@cunef.edu

Finance Cunef Madrid, Spain

References

[1] Agarwal, V, Daniel, N. D, & Naik, N. Y. Do hedge funds manage their reported returns?" *Drexel University, Georgia State University and CFR and London Business School.*

[2] Brooks Chris, and Kat, H.M. ((2002). The Statistical Properties of Hedge Fund Index Returns and their Implications for Investors." *Journal of Alternative Investments,* , 5

[3] Kat, H. M, & Miffre, J. (2006). The Impact of Non-Normality Risks and Tactical Trading on Hedge Fund Alphas." *Cass Business School, City University London Faculty of Finance Working Paper Series WP.*

[4] López Pascual J and Cuellar, R.D. ((2007). The challenges of launching, rating and regulating funds of hedge funds" *Journal of Derivatives & Hedge Funds, 133*

[5] López Pascual J and Cuellar, R.D. ((2008). Can the Hedge Fund regulation limit the negative impact of a systemic crisis" *Universia Business Review / Cuarto Trimestre 2008.*

[6] López Pascual J and Cuellar, R.D. ((2010). Assessment and Selection of Hedge Funds and Funds of Hedge Funds" Working Paper Nº5 2010, Cunef.

[7] López de Prado M.M. ((2008). How long should a track record be? *Journal of Alternative Investments*

[8] Mitchell Mark, Pedersen Lasse, Heje and Pulvino, Todd ((2007). Slow Moving Capital"

[9] Mitchell Mark, and Pulvino, Todd ((2001). Characteristics of Risk and Return in Risk Arbitrage" Journal of Finance, 56.

Permissions

The contributors of this book come from diverse backgrounds, making this book a truly international effort. This book will bring forth new frontiers with its revolutionizing research information and detailed analysis of the nascent developments around the world.

We would like to thank Fausto Pedro García Márquez and Benjamin Lev, for lending their expertise to make the book truly unique. They have played a crucial role in the development of this book. Without their invaluable contribution this book wouldn't have been possible. They have made vital efforts to compile up to date information on the varied aspects of this subject to make this book a valuable addition to the collection of many professionals and students.

This book was conceptualized with the vision of imparting up-to-date information and advanced data in this field. To ensure the same, a matchless editorial board was set up. Every individual on the board went through rigorous rounds of assessment to prove their worth. After which they invested a large part of their time researching and compiling the most relevant data for our readers. Conferences and sessions were held from time to time between the editorial board and the contributing authors to present the data in the most comprehensible form. The editorial team has worked tirelessly to provide valuable and valid information to help people across the globe.

Every chapter published in this book has been scrutinized by our experts. Their significance has been extensively debated. The topics covered herein carry significant findings which will fuel the growth of the discipline. They may even be implemented as practical applications or may be referred to as a beginning point for another development. Chapters in this book were first published by InTech; hereby published with permission under the Creative Commons Attribution License or equivalent.

The editorial board has been involved in producing this book since its inception. They have spent rigorous hours researching and exploring the diverse topics which have resulted in the successful publishing of this book. They have passed on their knowledge of decades through this book. To expedite this challenging task, the publisher supported the team at every step. A small team of assistant editors was also appointed to further simplify the editing procedure and attain best results for the readers.

Our editorial team has been hand-picked from every corner of the world. Their multi-ethnicity adds dynamic inputs to the discussions which result in innovative outcomes. These outcomes are then further discussed with the researchers and contributors who give their valuable feedback and opinion regarding the same. The feedback is then collaborated with the researches and they are edited in a comprehensive manner to aid the understanding of the subject.

Apart from the editorial board, the designing team has also invested a significant amount of their time in understanding the subject and creating the most relevant covers. They scrutinized every image to scout for the most suitable representation of the subject and create an appropriate cover for the book.

The publishing team has been involved in this book since its early stages. They were actively engaged in every process, be it collecting the data, connecting with the contributors or procuring relevant information. The team has been an ardent support to the editorial, designing and production team. Their endless efforts to recruit the best for this project, has resulted in the accomplishment of this book. They are a veteran in the field of academics and their pool of knowledge is as vast as their experience in printing. Their expertise and guidance has proved useful at every step. Their uncompromising quality standards have made this book an exceptional effort. Their encouragement from time to time has been an inspiration for everyone.

The publisher and the editorial board hope that this book will prove to be a valuable piece of knowledge for researchers, students, practitioners and scholars across the globe.

List of Contributors

Fausto Pedro García Márquez and Marta Ramos Martín Nieto
Ingenium Research Group, Universidad Castilla-La Mancha, Ciudad Real, Spain

William P. Fox
Naval Postgraduate School, Monterey, California, USA

Yi Liao
Southwestern University of Finance and Economics, China

Wenjing Shen and Benjamin Lev
Drexel University, USA

Xinxin Hu
Indiana University, USA

Margaret O. Afolabi and Omoniyi Joseph Ola-Olorun
Department of Clinical Pharmacy and Pharmacy Administration, Faculty of Pharmacy, Obafemi Awolowo University, Ile-Ife, Nigeria

Mahelet G. Fikru
Department of Economics, Missouri University of Science and Technology, USA

José Ignacio Muñoz Hernández
University of Castilla-La Mancha UCLM, Spain

José Ramón Otegui Olaso and Alejandro Gutiérrez López
University of the Basque Country UPV/EHU, Spain

José Ignacio Muñoz Hernández
University of Castilla-La Mancha UCLM, Spain

José Ramón Otegui Olaso
University of the Basque Country UPV/EHU, Spain

Julen Rubio Gómez
Dresser-Rand Inc., U.S.A.

William P. Fox
Naval Postgraduate School, USA

Fausto P. Garcia
Universidad Castilla-La Mancha, Spain

Joaquín López Pascual
Finance Cunef Madrid, Spain

Printed in the USA
CPSIA information can be obtained
at www.ICGtesting.com
JSHW011423221024
72173JS00004B/648